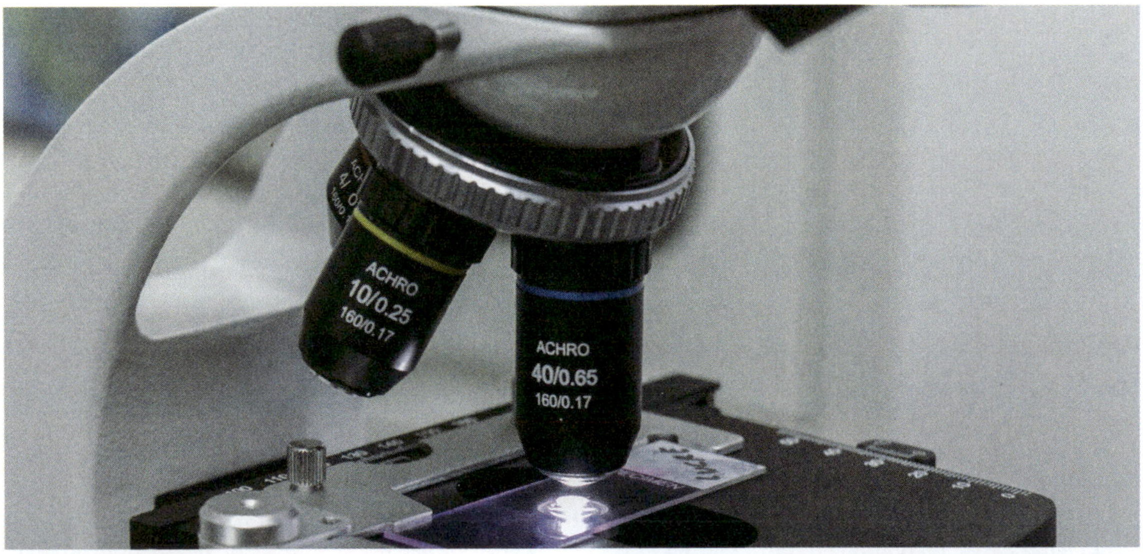

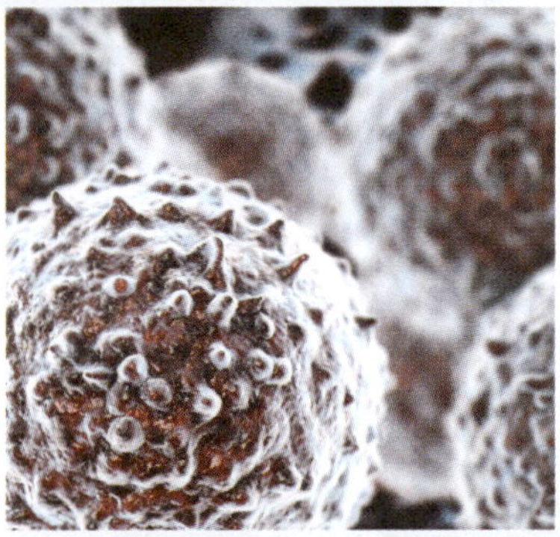

Principles of Biology Laboratory

BIOLOGY 116

Fifth Edition | West Virginia University

Katrina Stewart | Cheryl Walton | Nicholas Haas | Nikki Byrne-Hoffman

VAN-GRINER LEARNING

Principles of Biology Laboratory
Biology 116

Fifth Edition
West Virginia University
Katrina Stewart, Cheryl Walton, Nicholas Haas, and Nikki Byrne-Hoffman

Printed in the United States of America
10 9 8 7 6 5 4 3 2 1
ISBN: 978-1-61740-885-4

Van-Griner Publishing
Cincinnati, Ohio
www.van-griner.com

President: Dreis Van Landuyt
Project Manager: Maria Walterbusch
Customer Care Lead: Lauren Wendel

Stewart 885-4 Su21
317186

VAN-GRINER LEARNING

Biology 116 Acknowledgments

Sincere thanks and appreciation go to all who have helped with the Principles of Biology Labs. For inspiration and ideas for laboratory activities, we thank Katy Ryan, Jennifer Stueckle, Michelle Withers, and Stephanie Young. For support and suggestions, we thank faculty and staff of the Department of Biology, particularly Sophia Zhang, Kevin Barry, Dana Heubert Lima, Pat Lutsie, John Navaratnam, Kristi Teagarden, and Wendy Sites. For their efforts toward good teaching practices, we thank the Graduate Teaching Assistants. Thank you!

Biology 116
Table of Contents

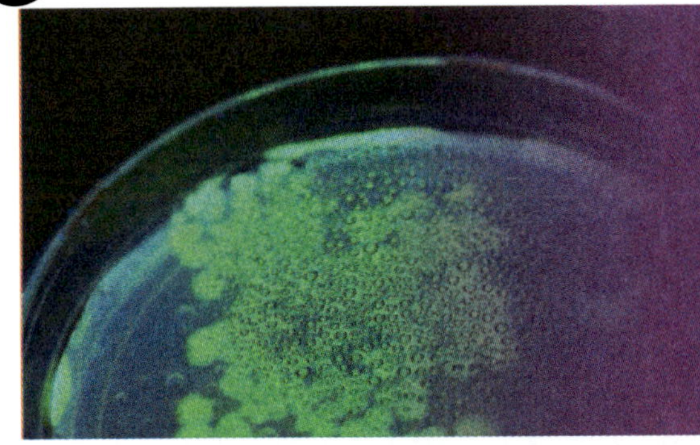

Biology 116
Fall 2021 Syllabus

TA:

Office:

Office Hours:

Phone:

Email:

Lab Managers: Cheryl Walton/
Nicholas Haas
Office: G2 Oglebay Hall
Office Hours: By appointment
Phone: (304) 293-6484
Email: biology116lab@mail.wvu.edu

Lab Managers: Nikki Byrne-Hoffman
Office: 208 NRCCE
Office Hours: By appointment
Phone: (304) 293-5684

Academic Specialist and Lab Supervisor: Katrina Stewart
Email: biology116lab@mail.wvu.edu

Downtown

Office: G2 Oglebay Hall
Phone: (304) 293-8670

Evansdale

Office: 208 NRCCE
Phone: (304) 293-5197

Course Description

BIOL 116 is the corequisite laboratory course associated with BIOL 115 lecture. This laboratory course emphasizes the proper understanding and use of the scientific method to design and perform biological experiments. Discipline-specific communication techniques, including scientific writing, are also emphasized.

You must be registered for both Biology 115 (lecture) and Biology 116 (lab) during the same semester. **Exception:** If you have already completed Biology 115 with a passing grade, you will be allowed to register for Biology 116 alone.

This course meets once per week, 2 hours per meeting (1 credit hour) and fulfills GEF 2B: Science & Technology.

Biology 116 has been designated as a SpeakWrite course by the Eberly College of Arts and Sciences. As part of the Eberly College's commitment to fostering effective communication skills, this course will:

- Emphasize informal and formal modes of communication.
- Teach discipline-specific communication techniques.
- Use a process-based approach to learning that provides opportunities for feedback and revision.
- Base a substantial portion of the final grade on successful spoken and/or written performance (~33%).

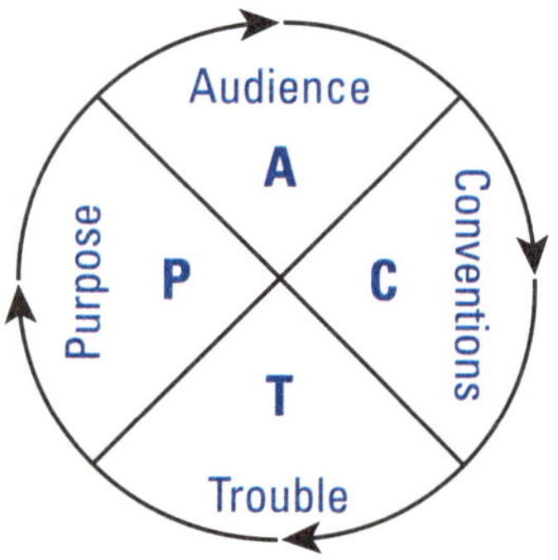

Purpose: What exactly do I want to happen?
Audience: Who is reading, listening, or viewing?
Conventions: What is expected in this context?
Trouble: What could get in the way of my goals?

Learning Objectives

Upon successful completion of the course, students will be expected to:

1. Work collaboratively to formulate hypotheses, design and run experiments, analyze and interpret data, and apply the scientific process to solve problems and make predictions.
2. Collect and record data that will be used to generate appropriate data tables and graphs.
3. Communicate experimental findings in a scientific format.
4. Understand and apply fundamental concepts in Cellular and Molecular Biology, Genetics, and Evolution.

Required Materials

Lab Manual: *Biology 116: Principles of Biology Laboratory Manual*

Course Website: Information about this course's policies, schedules, assignments, etc., can be accessed on the class eCampus website. Log on at: *https://eCampus.wvu.edu.*

Laptop, Computer, or Tablet: Needed for accessing the course website. Should courses be moved online, a webcam and microphone will also be required.

Safety Glasses and Attire: Safety glasses or goggles that meet ANSI Z87 standards are acceptable. Students must wear clothing that is safe for a laboratory setting (see Lab Safety).

Recommended Materials

Given that this is a writing intensive course, many students will find a guide to writing in biology useful. The following book is recommended, but not required.

Pechenik JA. 2016. A short guide to writing about biology. 9th ed. Boston (MA): Pearson Education.

Attendance Policies

Students are required to attend all labs. You are expected to attend and participate in the lab for the full amount of time scheduled for your section. Attendance is critical to your success in the course. In lab you will be collecting data from your experiments and receiving instructions needed to complete homework assignments.

1. Sign In

   You must sign your name on the attendance sheet to be considered present. Your signature is a record of your attendance. You may not sign in for students other than yourself.

2. Making Up Labs and Absences

   If you miss lab, **you must contact your TA within 24 hours** to schedule a make-up. It may be possible to make arrangements to make up a missed lab; however, you will only be allowed this accommodation **twice. Make-up labs are granted at the discretion of your TA and lab course coordinators.** If you miss the first day of class and do not make it up, it is considered an absence.

   Should you miss a lab, it is your responsibility to request and obtain information regarding any missed assignments from your TA. Your TA may instruct you to obtain information from your lab partners. No deadline adjustments will be made for assignments, so please gather information and data in a timely manner to allow yourself adequate time to complete your assignment.

   If you choose not to make up a lab, you will be marked absent and will not earn points for participation or in-class work. **More than two absences will result in a grade of F for the lab course.**

3. Tardy Policy

   If you arrive late to lab, your TA will record your tardiness. If you arrive late a second time, you will not earn points for participation or in-class work. Alternatively, your TA may ask you to leave the lab and make it up at another time. If you have already made up two labs when you arrive late, you may be allowed to participate for the remainder of class, but will not receive credit for in-class work.

4. Serious Circumstances

   Students who miss classes because of extreme circumstances such as an extended hospital stay or death in the family should contact the office of Student Life in E. Moore Hall at (304) 293-5811 (*https://studentlife.wvu.edu/*) as soon as possible. At their discretion, this office may provide letters to inform a student's instructors "under certain limited conditions." For typical illnesses and absences, follow the syllabus policies outlined above.

5. University Excusable Absences (e.g., athletics)

   Students who expect to have University excusable absences (e.g., athletes, members of the Pride) should discuss absences with their TA as soon as possible at the beginning of the semester and create a plan for scheduled missed labs.

Grading

***You must be registered for both Biology 115 (lecture) and Biology 116 (lab) during the same semester.* Exception:** If you have already completed Biology 115 with a passing grade, you will be allowed to register for Biology 116 alone. Your grades for BIOL 115 and 116 are not combined.

Your grade will be based on your performance on the following:

Online Quizzes: Weekly quizzes will test for basic understanding and application of concepts. They will include questions from completed labs as well as questions that require you to explore outside sources of information to prepare for upcoming labs. Quizzes assigned each week are due prior to the following lab.

In-Class Activities and Assignments: You will be working in class as a group to complete such tasks as brainstorming ideas, solving problems, and collecting experimental data. You will be asked to submit documentation of your work. For example, following some sessions you may be asked to submit a copy of lab notes. After other sessions you may be asked to submit a worksheet that shows calculations of practice data sets.

Homework Assignments: These assignments are to be completed outside of class and may include worksheets, summaries of scientific articles, peer reviews of students' work, and lab reports. Lab reports are written in the style of a peer-reviewed primary scientific journal article and contain introduction, methods, results with figures and/or tables, and discussion sections. Each student is expected to perform library research and find information from textbooks, review articles, and primary journal articles to cite in their reports. You will synthesize knowledge gained from these to provide background information and rationale for each of the experiments performed in the lab, then compare and contrast your own experimental findings with the published data. Figures and tables are included in all reports, and results from appropriate statistical analyses are used draw conclusions. You will receive comments and feedback on drafts of reports before a final version is submitted to allow you the opportunity to refine writing styles. Guidelines and a rubric for lab reports will be provided.

You are encouraged to make use of the on-campus resources to help with your writing. For handouts on getting started with writing up to revising and editing go to *http://speakwrite.wvu.edu*. For more individualized help, the Eberly Writing Studio in G02 Colson Hall has trained peer consultants that will work with you. To make an appointment call (304) 293-5788 or visit the Eberly Writing Studio link on the SpeakWrite website.

Assignments

Assignments are due at the beginning of the class period on the specified due date; 10% will be deducted for each day the assignment is late, including weekends. Late work will not be accepted after 10 days. You must submit the assignments by the beginning of lab in order for them to be considered on time even if you do not attend the lab the day the assignment is due.

1. Submitting Assignments

   Most assignments will require electronic submissions through Turnitin links on eCampus. Emailed assignments will not be graded. **If there are difficulties submitting assignments through Turnitin links on eCampus, you must submit a ticket to the ITS help desk to resolve the issue (*https://wvu.teamdynamix.com/TDClient/Home/*). You must also email the assignment and associated help desk ticket number to your TA by the due date and time to show it was completed on time.** Your assignment will not be graded until it is properly submitted through Turnitin.

   Carefully follow instructions and guidelines when submitting assignments through Turnitin links in eCampus to ensure that your homework is submitted properly. Your assignment must be in a readable format that your TA can grade. It must also be in a format that can be analyzed for originality by Turnitin software. No extensions or exceptions will be given for assignments that do not meet these requirements. Submit Microsoft Word (.doc) or PDF (.pdf) files. Documents that are scanned documents or image files are not acceptable. If you use Apple Pages or Google Docs, you must convert the files to Microsoft Word or PDF files prior to submitting. Late policies apply for any submissions that do not meet these requirements by the due date.

   After submitting documents:

   - Immediately open the assignment and make sure that it uploaded.
   - Check that it is the correct assignment for the Turnitin dropbox.
   - Check that the document is in an appropriate format (.doc or .pdf).
   - Check that the document text is visible in the Feedback Studio window. If a bar at the top of the window is prompting to download, it has **not** uploaded properly and cannot be graded. Check the file and re-upload or ask for assistance.

2. eCampus Gradebook

   Assignment scores will be posted on eCampus throughout the semester. It is your responsibility to monitor your scores. Any questions regarding scores must be brought to the attention of your TA and/or lab course coordinators within two weeks of the posting date for individual assignments. After this time period, requests to review scores will not be considered.

Academic Integrity

The integrity of the classes offered by any academic institution solidifies the foundation of its mission and cannot be sacrificed to expediency, ignorance, or blatant fraud. Therefore, TAs and course coordinators will enforce rigorous standards of academic integrity in all aspects and assignments of this course. For the detailed policy of West Virginia University regarding the definitions of acts considered to fall under academic dishonesty and possible ensuing sanctions, please see the Student Conduct Code *http://studentconduct.wvu.edu/home*. Should you have any questions about possibly improper research citations or references, or any other activity that may be interpreted as an attempt at academic dishonesty, please see your TA *before* the assignment is due to discuss the matter.

Assignments completed outside of lab are expected to be done individually and represent independent work. Acts of academic dishonesty or plagiarism of any sort will not be tolerated and will be reported immediately according to University policies and procedures. Examples of unacceptable conduct include, but are not limited to:

- Copying another student's work.
- Improperly citing sources.
- Working collaboratively on independent assignments.
- Providing your work for another student to use, either directly or via posting online.
- Presenting another author's words as your own.
- Fabricating data.

Inclusivity Statement

The West Virginia University community is committed to creating and fostering a positive learning and working environment based on open communication, mutual respect, and inclusion.

If you are a person with a disability and anticipate needing any type of accommodation in order to participate in your classes, please advise your instructors and make appropriate arrangements with the **Office of Accessibility Services.** (*https://accessibilityservices.wvu.edu/*)

More information is available at the **Division of Diversity, Equity, and Inclusion** (*https://diversity.wvu.edu/*) as well.

Sale of Course Material Statement

All course materials, including lectures, class notes, quizzes, exams, handouts, presentations, and other course materials provided to students for their courses are protected intellectual property. As such, the unauthorized purchase or sale of these materials may result in disciplinary sanctions under the Student Conduct Code. *(https://studentconduct.wvu.edu/campus-student-code)*

Appropriate Use of Technology Statement

Use of technology in the classroom should always be directly related to class activities and/or course learning outcomes. Inappropriate technology use can be an impediment to learning and a distraction to all members of the class. As such, inappropriate use of technology in the classroom may be considered a disruption of the class and constitute a violation of the **WVU Student Conduct Code** and could potentially result in a referral to the Office of Student Conduct. Use of technology in the classroom when specifically prohibited by the instructor may also constitute a violation of WVU's **Academic Integrity** policy.

WVU Course Policies

For the policies that govern all WVU courses, please see *https://tlcommons.wvu.edu/syllabus-policies-and-statements.* In addition to the policies and statements above, pay special attention to:

- Academic Standards Policy, including Academic Dishonesty
- Adverse Weather Statement
- Attendance Policy
- COVID-19 Syllabus Statement

Lab Safety

1. **Dress Policy.** Close-toed shoes, long pants/skirts that come below the knee when standing and sitting, and covered shoulders and midriff are required at all times while working in the laboratory. No shorts or short skirts are allowed. Ripped jeans or pants with holes that expose skin are not acceptable. You may not be permitted to attend lab if you fail to dress appropriately.

2. **Safety glasses or goggles are to be worn in the lab.** Many experiments present hazards, so safety glasses/goggles that meet ANSI Z87 standards must be worn. In addition, students must adhere to instructions for wearing gloves and following all safety protocols. Due to close working arrangements, masks are strongly encouraged to reduce exposure to transmissible illnesses.

3. **Please do not enter the room until TA is present.** The TA is responsible for all equipment in the lab, as well as your health and welfare during the class period. Wait in the hallways until your TA arrives.

4. **Do not use cell phones, smartwatches, or other electronic devices without permission from your TA.** No texting, or placing/receiving voice calls is permitted during lab. Cell phones are not allowed on lab benches. They should be stored in bags/backpacks and out of sight during lab unless an exception is made for class activities.

5. **No smoking/tobacco use, vaping, eating, drinking, or chewing gum in the laboratory.**

6. **Wash hands before you leave lab for the day.**

7. **Make sure your work area is clean before you begin, and clean your area and all labware after you have finished for the day.**

8. **Any accidents should be reported promptly to your instructor.**

9. **Make sure you know the location of building exits.** Do not use the elevator during an emergency.

   - **Emergency Exit Routes:** Please follow your TA's instructions on which exit to use so that he or she can account for students in the class and determine if all are safely out of the building.

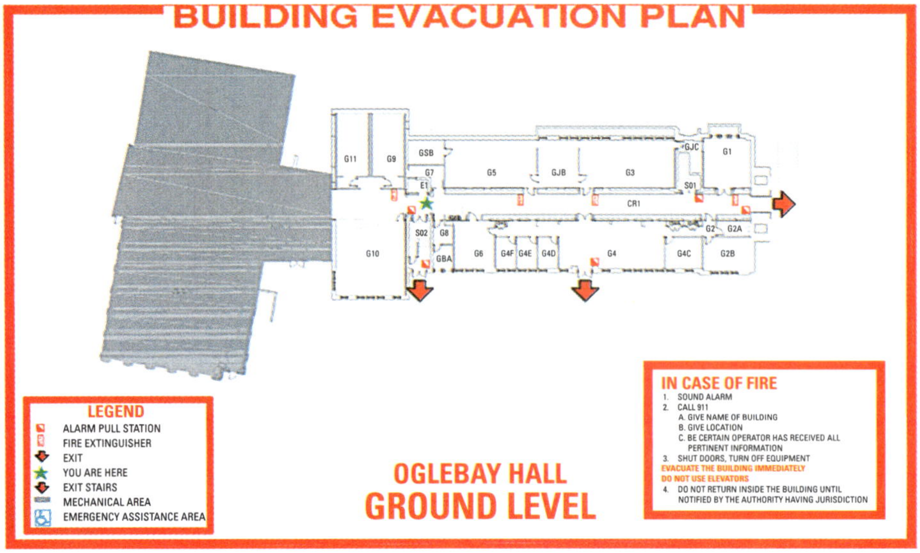
BUILDING EVACUATION PLAN
G11
G9
GSB
G7
E1
G5
GJB
G3
GJC
G1
S01
CR1
G10
S02
G8
GBA
G6
G4F
G4E
G4D
G4
G2
G2A
G4C
G2B
LEGEND
ALARM PULL STATION
FIRE EXTINGUISHER
EXIT
YOU ARE HERE
EXIT STAIRS
MECHANICAL AREA
EMERGENCY ASSISTANCE AREA
OGLEBAY HALL
GROUND LEVEL
IN CASE OF FIRE
1. SOUND ALARM
2. CALL 911
A. GIVE NAME OF BUILDING
B. GIVE LOCATION
C. BE CERTAIN OPERATOR HAS RECEIVED ALL PERTINENT INFORMATION
3. SHUT DOORS, TURN OFF EQUIPMENT
EVACUATE THE BUILDING IMMEDIATELY
DO NOT USE ELEVATORS
4. DO NOT RETURN INSIDE THE BUILDING UNTIL NOTIFIED BY THE AUTHORITY HAVING JURISDICTION

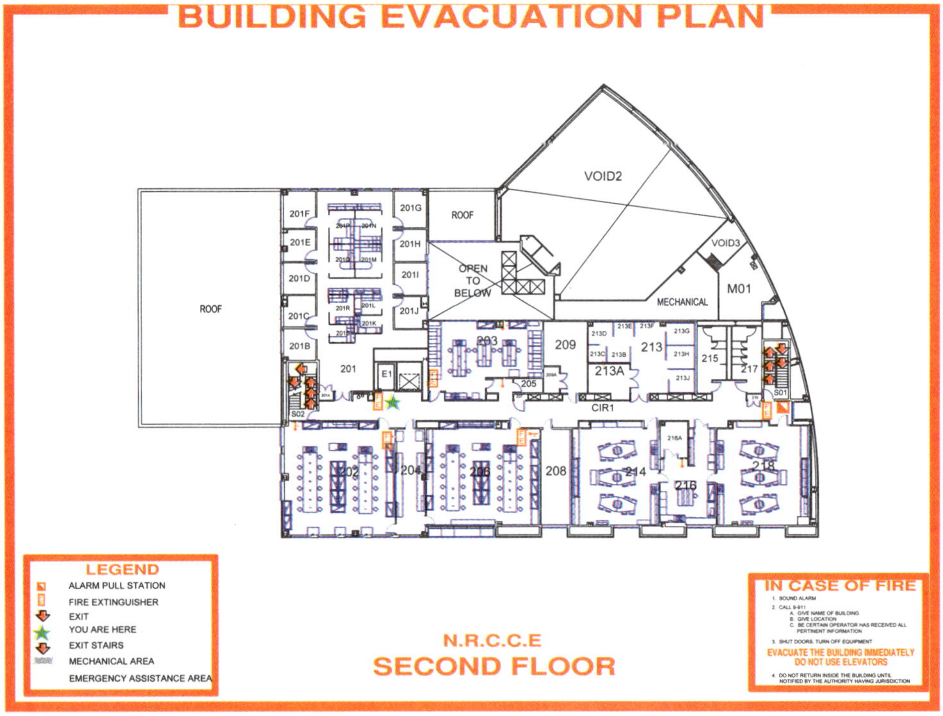
BUILDING EVACUATION PLAN
VOID2
VOID3
ROOF
201F
201E
201D
201C
201B
201G
201H
201I
201J
ROOF
OPEN TO BELOW
MECHANICAL
M01
201
E1
203
205
209
213
213A
215
217
S01
S02
CIR1
202
204
206
208
214
216
218
LEGEND
ALARM PULL STATION
FIRE EXTINGUISHER
EXIT
YOU ARE HERE
EXIT STAIRS
MECHANICAL AREA
EMERGENCY ASSISTANCE AREA
N.R.C.C.E
SECOND FLOOR
IN CASE OF FIRE
EVACUATE THE BUILDING IMMEDIATELY
DO NOT USE ELEVATORS

Biology 116
Make-Up Lab Form 1

Instructions to Students: You must fill out the top portion of the form and ask the TA from the make-up lab to fill out the bottom portions of the form. You must then take the original and give it to your regular TA so your make-up lab can be recorded.

It is your responsibility to give the form to your TA. Keep a copy for your records.

I have attended a make-up lab session. I have participated in the lab activity and will contact my TA for information I may have missed in my regular section.

Student's Name: ______________________________

Student's Signature: ______________________ Date: __________

Section #: ______ Day of the Week M/T/W/R/F Time: ______________

Name of Regular Lab TA: ______________________________

This student has attended my lab session and participated in the lab activity.

Make-Up Lab Section #: _____ Day of the Week M/T/W/R/F Time: __________

In-Class Points Earned: __________ Comments: ____________________

____

Name of TA for Make-Up Lab: ______________________________

Signature of TA for Make-Up Lab: ____________________ Date: __________

Biology 116 Make-Up Lab Form 1

Instructions to Students: You must fill out the top portion of the form and ask the TA from the make-up lab to fill out the bottom portions of the form. You must then take the original and give it to your regular TA so your make-up lab can be recorded.

It is your responsibility to give the form to your TA. Keep a copy for your records.

I have attended a make-up lab session. I have participated in the lab activity and will contact my TA for information I may have missed in my regular section.

Student's Name: ____

Student's Signature: ______________________________ Date: ____________

Section #: ________ Day of the Week M/T/W/R/F Time: ____________________

Name of Regular Lab TA: ______________________________________

This student has attended my lab session and participated in the lab activity.

Make-Up Lab Section #: ______ Day of the Week M/T/W/R/F Time: ____________

In-Class Points Earned: ____________ Comments: ____________________________

____

Name of TA for Make-Up Lab: ___________________________________

Signature of TA for Make-Up Lab: ____________________ Date: ____________

Biology 116
Make-Up Lab Form 2

Instructions to Students: You must fill out the top portion of the form and ask the TA from the make-up lab to fill out the bottom portions of the form. You must then take the original and give it to your regular TA so your make-up lab can be recorded.

It is your responsibility to give the form to your TA. Keep a copy for your records.

I have attended a make-up lab session. I have participated in the lab activity and will contact my TA for information I may have missed in my regular section.

Student's Name: ______________________________

Student's Signature: ______________________ Date: __________

Section #: ______ Day of the Week M/T/W/R/F Time: ______________

Name of Regular Lab TA: ______________________________

This student has attended my lab session and participated in the lab activity.

Make-Up Lab Section #: _____ Day of the Week M/T/W/R/F Time: __________

In-Class Points Earned: __________ Comments: ____________________

____

Name of TA for Make-Up Lab: ______________________________

Signature of TA for Make-Up Lab: ____________________ Date: __________

Biology 116 Make-Up Lab Form 2

Instructions to Students: You must fill out the top portion of the form and ask the TA from the make-up lab to fill out the bottom portions of the form. You must then take the original and give it to your regular TA so your make-up lab can be recorded.

It is your responsibility to give the form to your TA. Keep a copy for your records.

I have attended a make-up lab session. I have participated in the lab activity and will contact my TA for information I may have missed in my regular section.

Student's Name: ______________________________

Student's Signature: ______________________ Date: __________

Section #: ______ Day of the Week M/T/W/R/F Time: ______________

Name of Regular Lab TA: ______________________________

This student has attended my lab session and participated in the lab activity.

Make-Up Lab Section #: ____ Day of the Week M/T/W/R/F Time: __________

In-Class Points Earned: ________ Comments: ____________________

____

Name of TA for Make-Up Lab: ______________________________

Signature of TA for Make-Up Lab: ________________ Date: __________

Lab 1 | Introduction to Lab

Assignments

In-Class

- ❐ Notes and participation

Homework

- ❐ Read Lab 2: How Do We Know?—Using the Scientific Process
    - ❐ Do the "Practice with Scientific Hypotheses" questions #1–10. Write your answers in your lab manual and bring your answers to the next lab.
    - ❐ Do the "Practice Identifying Part of Experiments" Experiment Example 1 and 2 in your lab manual and bring your answers to the next lab.
- ❐ Quiz

Learning Goals

By the end of this lab, you will be able to:

- Explain the format of the class, including course learning goals and objectives
- Apply good lab safety practices, prevent accidents, and be prepared to respond if one should occur.
- Collaborate with other people and communicate effectively with group members

Introduction

Biology 116 laboratory is an introductory course focused on teaching basic lab skills, critical thinking, and scientific writing. This course was designed to mimic a real research laboratory. Over the next several weeks, you will be introduced to scientific literature and basic lab skills. You will learn how to develop a hypothesis based on observations, design experiments to test hypotheses, and communicate your findings in a scientific manner.

Your Teaching Assistant

At the top of your course syllabus, fill out the information with your TA's name, contact information, and office hours. You will need this information throughout the semester if you have any questions regarding homework or make-up labs.

Your Group Members

You will be working with group members in lab throughout the semester. Today, in addition to getting to know your TA, you should get to know the members of your group. Fill out your name card as directed by your TA.

Figure 1 Name card.

Introduce yourself to your group members, exchange contact information, and decide on a "team" name. When finished, your TA will lead a discussion about group dynamics.

Always remember when working in groups:

- Keep it safe
- Keep it positive and constructive
- Everyone gets a voice
- Everyone participates/contributes

Constructive and Destructive Group Behaviors

Read the list of constructive and destructive behaviors people exhibit when working in groups. Select a behavior from each list with which you most identify and discuss them with the group. How will your group take advantage of the constructive behaviors? How will you be able to manage destructive behaviors?

Constructive Group Behaviors

- **Cooperating:** Is interested in the views and perspectives of the other group members and is willing to adapt for the good of the group.
- **Clarifying:** Makes issues clear for the group by listening, summarizing, and focusing discussions.
- **Inspiring:** Enlivens the group, encourages participation and progress.
- **Harmonizing:** Encourages group cohesion and collaboration. For example, uses humor as a relief after a particularly difficult discussion.
- **Risk Taking:** Is willing to risk possible personal loss or embarrassment for the group or project success.
- **Process Checking:** Questions the group on process issues such as agenda, time frames, discussion topics, decision methods, use of information, etc.

Destructive Group Behaviors

- **Dominating:** Takes much of meeting time expressing self views and opinions. Tries to take control by use of power, time, etc.
- **Rushing:** Encourages the group to move on before task is complete. Gets "tired" of listening to others and working as a group.
- **Withdrawing:** Removes self from discussions or decision making. Refuses to participate.
- **Discounting:** Disregards or minimizes group or individual ideas or suggestions. Severe discounting behavior includes insults, which are often in the form of jokes.
- **Digressing:** Rambles, tells stories, and takes group away from primary purpose.
- **Blocking:** Impedes group progress by obstructing all ideas and suggestions. "That will never work because …"

Adapted from Brunt (1993). *Facilitation Skills for Quality Improvement. Quality Enhancement Strategies,* Madison WI. Summer Institute on Undergraduate Education in Biology 2008 Scientific Teaching: Diversity, Assessment, Active Learning Facilitator Guide. Page 10.

Lab 2 | How Do We Know?—Using the Scientific Process

Assignments

In-Class

- ❒ Notes and participation

Homework

- ❒ Complete a Testing Hypotheses Worksheet based on the experiment you design and carry out. Submit the assignment using Turnitin on eCampus
- ❒ Read Lab 3: Practicing Lab Skills—Just do it!
- ❒ Quiz

Learning Goals

By the end of this lab, you will be able to:

- Make and record observations
- Present unique questions about the topic
- State complete scientific hypotheses for your questions
- Design and carry out experiments that are appropriate to test the hypotheses:
  - Incorporate proper constants and controls
  - Identify independent and dependent variables
  - Include adequate replication
  - Record data
- Present summarized data using graphs
- State whether or not your hypothesis was supported using the data you collect

Introduction

Have you ever wondered how we know antibiotics are effective in treating infections? Or how we know the age of a newly found fossil? Or how we know certain genes have specific functions? Maybe you've wondered how salmon can navigate back to their home streams to spawn after traveling thousands of miles out to sea. Or why chemotherapies are usually more effective in treating blood cancers, like leukemia, than carcinoid tumors of organs.

Scientists ask questions like these every day. To find the answers, they use the scientific process. This means they are making observations, asking questions, formulating hypotheses, conducting experiments, analyzing data, and drawing conclusions.

Making Observations and Asking Questions—The Beginning of Inquiry

As you look at the world around you, you may notice patterns or maybe you see something unusual. These **observations** can lead you to start to ask "why" or "how" **questions** about what you are seeing—not unlike what scientists do in their work. Scientists may make observations during their own research as well as when they read through the scientific literature, taking notice of what others have found. Results from completed research become observations that can lead to the next questions. Regardless of this non-linear, cyclical nature of science, all inquiries begin with making observations and posing good questions. The next step is to begin looking for answers.

Hypotheses—Beyond Guessing

You have probably heard of **hypotheses** and may think of them as "educated guesses." However, hypotheses are much more than educated guesses—even if the "guess" is based on prior knowledge or observations. *Hypotheses are proposed explanations for an unexplained occurrence; they are possible solutions to a scientific question.*

Beyond being a proposed explanation, good scientific hypotheses must be testable and falsifiable. In other words, you must be able to design an experiment and collect data that will either support or refute the hypothesis. In addition, you need to be able to make predictions related to your hypothesis; you'll need to be able to say what you expect to happen if the hypothesis is true.

Hypotheses are sometimes structured as if-and-then statements. For students wanting to make sure their hypotheses are complete, the if-and-then format is a good model. It goes like this:

IF (proposed explanation, hypothesis), AND (experiment, test), THEN (prediction).

Notice how all of the essential parts are here. Not only is the hypothesis properly presented, but also there is evidence that it is testable, falsifiable, and that a prediction can be made. Consider the example:

> Based on the observation that we lose function in our hands when they are cold, students were asked to present hypotheses to explain this phenomenon. Here's what they came up with and why each example is or isn't a complete scientific hypothesis.

Question: Can temperatures affect hand function?

Student "hypotheses":

a. I can break more toothpicks with my hand when it is warm than I can when my hand is cold.

No, not a scientific hypothesis. This is only a prediction; there is no possible explanation or way to test it.

b. If I break toothpicks for one minute with my warm hand and then with my cold hand, then I will break more toothpicks with my hand when it is warm.

No, not a scientific hypothesis. This is only a method/test and prediction; there is still not a possible explanation included.

c. If low temperatures suppress muscle contractions and thus fine motor skills, and I break toothpicks for one minute with my warm hand and then for one minute with my hand after soaking it in ice water for five minutes, then I will break more toothpicks with my hand when it is warm.

Yes, this is a scientific hypothesis. There is a possible explanation, an experiment or a way to test it, and a predicted outcome of the experiment

Practice with Scientific Hypotheses

Look at the following examples of "hypotheses." Read each one and look for the important components of a complete scientific hypothesis. They might not be in the if-and-then format, but you should still be able to pick out and identify the following:

A. Proposed answer to a question (identify this part by underlining it)

B. Testability/falsifiability (identify this part by drawing a box around it)

C. A prediction (identify this part by circling it)

In addition to identifying parts, indicate which (if any) parts are missing.

1. Monkeys will eat apples before bananas. Missing A & B

2. If chemicals found in industrial pollution disrupts reproduction in seahorses, and you compare seahorses in polluted and unpolluted waters, then seahorses in polluted waters will produce fewer gametes (eggs and sperm).

3. Bacterial cells with multiple genes coding for antibiotic resistance will likely have better survival rates when exposed to new antibiotics.

Missing B

Missing A & B

4. Cats and other predators are the cause of songbird extinction on islands.

5. If fungal infections are measured in pet cats that live outdoors, the number of fungal infections in their owners will be higher. Missing B

6. The enzymes for chlorophyll production are necessary for plant maturation. When life cycles of albino plants and green plants are compared, only green plants will reach maturation.

7. If climate change is responsible for rising sea levels, sea levels will be different.

Missing B

8. Spirulina, blue green algae, will lower blood lipids in people that consume it.

9. If HIV replication can be prevented with CRISPR-Cas9 gene editing technology, HIV levels can be measured in mice treated with CRISPR-Cas9 and mice that are not. Missing C

10. If fish populations are only adapted to thrive in their local environments, and you swap fish from a lake and an adjacent stream into the opposite environments and monitor fish fitness, non-native populations will be less fit that the native populations.

Designing the Experiment—Putting Your Hypothesis to the "Test"

Recall that a good hypothesis is testable—that often means setting up an experiment. If you expect to collect data that enables you to support or refute your hypothesis, then you need a properly designed experiment. This usually means having **experimental groups** and a control or a **control group.** For example, suppose medical researchers were testing the effectiveness of a newly developed drug. Patients who receive the drug treatment would be part of an experimental group. A control group is used as a baseline for comparison with experimental groups. Patients who receive no drug or a placebo would be part of the control group. The experimental groups will differ from the control by a single factor that you choose—one variable that you will manipulate. This manipulated variable is called the **independent variable,** and in this example would be the drug treatment. The effects of the independent variable can be measured—these measurements are called **dependent variables.** Researchers studying the new drug may choose to measure the effectiveness in a variety of ways depending on their specific question. Their dependent variables might include severity of symptoms or maybe time needed for patients to experience relief from symptoms.

You can only manipulate one factor at a time, so everything else among your treatments and control groups must stay the same. If they don't, you won't know if your results are due to your independent variable or something else. The **constants** in your experiment are all of those factors that you keep the same. Sometimes constants are called controlled variables, but be careful not to confuse constants with controls or control groups. Researchers in the new drug example would be sure to establish and monitor constants. They would likely give equivalent doses of drugs to patients, choose patients for the study that didn't have physical conditions that might interfere with the new drug or measurement of symptoms, and they may even choose to administer the drug in a controlled environment.

Practice Identifying Parts of an Experiments

In the following examples, identify the independent variables, dependent variables, control groups, and constants.

Experiment Example 1

Grasses have tillers, which are stems that grow from the base of the plants. More tillers mean more biomass, which can be converted into biofuels. Domesticated grass plants have been selected to produce more grain instead of tillers. However, Dr. Jennifer Hawkins at the WVU Department of Biology isn't interested in more grain—she is interested in plants with more tillers. More specifically, she wants to know what genes could be genetically engineered to produce more tillers to increase plant biomass production. By comparing the genomes of wild ancestors with more tillers to domesticated plants, she identifies genes that could be modified to enhance tiller growth. In one of her experiments, *Sorghum* spp. plants were grown in a greenhouse under carefully monitored conditions, such as light and humidity. By increasing biomass through more tillers, genetic engineering could help reduce the production costs of grass plants used to produce biofuels.

- Independent variable(s): The types of genes in the plant
- Dependent variable(s): How much till is produced from the grass
- Control(s): light and humidity
- Constants: Grown in a greenhouse.

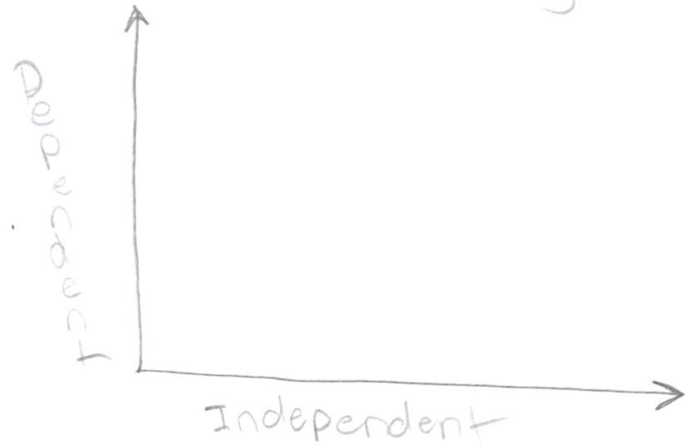

Experiment Example 2

Triclosan is an antimicrobial chemical incorporated into many skin care, medical, and household products, including hand soaps. Due to health concerns related to the overuse of antimicrobial products, the Federal Drug Administration (FDA) has recently banned the use of such chemicals in most over-the-counter personal care products. Shortly before the FDA ban, researchers at the National Institute for Occupational Safety and Health, part of the Centers for Disease Control, in Morgantown, WV, studied whether triclosan promotes an allergic response on the skin. They applied a solution containing triclosan to the skin of mice and compared their allergic responses to responses of mice treated with a control solution. As expected, the mice with the triclosan treatments showed increased allergic responses. The mice used in the experiment were a laboratory-bred strain commonly used in experiments because they lack immune responses to many treatments, so any responses to triclosan were easily observed.

- Independent variable(s): The amount of solution put on the mice
- Dependent variable(s): Allergic reactions caused by the solution
- Control(s): The enviroment the mice are kept in
- Constants: Bread of mice

Replication—How Many Samples or Measurements Should You Collect?

Determining how many **replicates** are needed (i.e., how many samples or measurements to take) can be tricky because there is no magic number that will work for all experiments universally. You don't want to waste valuable time and resources collecting too many unneeded samples. However, you should be wary of collecting too few samples because you could end up unable to draw meaningful conclusions from your data. Fewer replicates increase the likelihood that your results are due to chance rather that your independent variable. In general, don't stop with the minimum if it is reasonable to collect more data—more replicates can usually strengthen your data set and make your results more reliable. In the previous example experiment about the new drug study, the researchers are more likely to be confident in the results if they had 1,000 patients than if they only had 10.

Making Conclusions—Supporting or Refuting the Hypothesis

It is important to note that while you can falsify a hypothesis, or prove it to be false, you can **never** prove a hypothesis to be true. The data you collect can make you very confident that your hypothesis is correct, but you cannot claim to be 100% certain. To illustrate this point, consider this example: Let's say you observe hummingbirds visiting flowers. You wonder what benefit, if any, does the plant receive from these visits. You hypothesize that hummingbirds help the plant reproduce by transferring pollen from flower to flower. You predict that if you monitor hummingbirds visiting flowers, you will see pollen stuck to hummingbird beaks and feathers. Suppose you monitor the hummingbirds and find that they do not have pollen stuck to their beaks and feathers; you could reject your hypothesis. However, what if you find that hummingbirds do in fact have pollen on their beaks and feathers? This would certainly support your hypothesis, but it could not prove it. There may be other reasons why the pollen is stuck to the bird. For example, maybe it is a source of food for the hummingbird. The message is this: *You can reject a hypothesis, or you can support a hypothesis, but you cannot ever prove a hypothesis to be true.*

Lab Activities

Now that you have practiced the basics generating hypotheses and recognize components of well-designed experiment, it is time to practice your skills.

Part 1: Do Termites Follow Trails?

You may have heard that termites will follow the ink trail of a pen. But have you ever seen it yourself? For this part of the lab, your task is to make observations about trail following behaviors of termites.

1. Using a ballpoint ink pen, draw a line or shape on a piece of paper to serve as a "trail" for a termite.
2. **Gently** place a termite on the paper near the trail using a paintbrush. Use the paintbrush to keep it on the paper if it starts to wander off.
   - Be careful not to break the termite's legs or otherwise harm it! Don't "sweep" it!
   - Placing the paintbrush in front of a termite and then rolling the brush away from it can encourage it to "jump" on.
3. Observe the termite for several minutes. Record your observations about the termite's behavior.

4. Continue to record your observations with different pens or pencils and termites. Be prepared to share your observations with the class.

5. Make a list of measurements you could use to quantify the termite behaviors. In other words, make a list of possible dependent variables you could use in an experiment. Share your ideas with the class.

Record your observations:

- The thermite walked back and forth on the straight line
- The thermite had trouble following the ink circle
- After losing the ink line it couldn't find it's way back without assistance

Part 2: Ask a Question and Generate a Hypothesis

Based on your observations, you likely have many questions about the termites' behaviors. Your tasks for this part of the lab are to decide on a question, then formulate a testable hypothesis.

Write your question and hypothesis:

- Thermites are able to track linear paths better than paths with twists and turns.

Part 3: Design and Carry Out Your Experiment

Now that you have a solid questions and testable hypotheses, you need an experiment.

1. Design your experiment to test your hypothesis and be sure to consider all of the important components, including independent variables, dependent variables (generate quantifiable data), constants, controls, and replication. You should also be sure to make sure your experiment is relevant to the original question.

2. Present your ideas for your experiment to your TA and/or the class. Once your TA gives you the okay, then you can begin.

3. Collect and record data.

4. Summarize your results by sketching figures (graphs, tables) to present to your TA and/or the class.

Detail your experimental design, record results, and sketch figures:

• The expirement is going to have three differant lines.

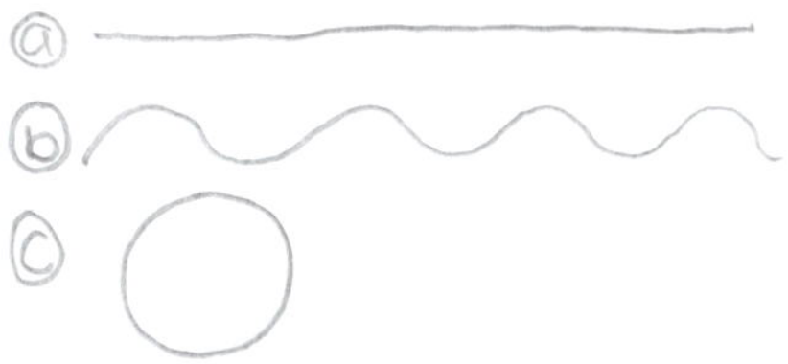

• We are then going to place a thermite on each line and track its trail.

• Indendent variable - The paths created by an ink pen

• Dependent Variable - The path the thermite decides to take.

• Constants - surface used, enviroment

• control - The thermite used

• The expirement will need to be done multiple times to collect accurate data.

Detail your experimental design, record results, and sketch figures (cont.):

Assignments

- In-Class
  - Take good notes in your lab manual. Actively participate in lab.
- Homework
  - Complete a Testing Hypotheses Worksheet using your lab manual notes. Type it using the electronic form for the assignment. Submit it through the Turnitin link on eCampus. This is an independent homework assignment. Expectations for this assignment will be discussed in lab, so please pay attention and take notice!
  - Read Lab 3: Practicing Lab Skills—Just do it!
  - Quiz

References

Johnson AD. 2009. 40 inquiry exercises for the college biology lab. Arlington (VA): NSTA Press.

Morris JR, Hartl DL, Knoll AH, Lue R. 2013. Biology: how life works. New York (NY): W.H. Freeman & Company.

Strode PK. 2015. Hypothesis generation in biology: a science teaching challenge & potential solution. Am Biol Teach. (777) 500–506.

Lab 3 | Practicing Lab Skills—Just Do It!

Assignments

In-Class

- ❒ Notes and participation

Homework

- ❒ Complete and submit the Lab Skills homework assignment.
- ❒ Read Lab 4: Investigating Enzymes—Mushroom Metabolism
- ❒ Quiz

Learning Goals

By the end of this lab, you will be able to:

- Measure small volumes accurately using micropipettes
- Perform dilution calculations
- Use a spectrophotometer to get absorbance values for solutions
- Create standard curves by hand and with Excel
- Quantify target molecules with a standard curve

Introduction

Today's laboratory exercise is designed to introduce you to different scientific tools that you will use in upcoming labs. You will practice using micropipettes for measuring small volumes of liquid, as well as use spectrophotometers for determining concentrations of substances in solution.

Concentration is defined as the amount of a certain molecule divided by the total volume of a mixture. For example, if 200 grams of glucose is added to one liter of water, then the concentration of glucose in the solution is 200 g/L. Concentrations are important to study in biological systems for many reasons. For example, the concentration of a solute can affect the rate of chemical reactions within an organism and can also affect diffusion across cell membranes.

During today's exercise you will find solution concentrations using a spectrophotometer. A spectrophotometer is an instrument that measures the amount of a specific wavelength of light that is absorbed by a solution. The amount of light absorbed is directly proportional to the concentration of molecules in the solution. (i.e., Solutions with a low concentration of a given molecule will absorb very little light and appear light in color, while solutions with a high concentration of a given molecule will absorb a great deal of light and appear darker in color.)

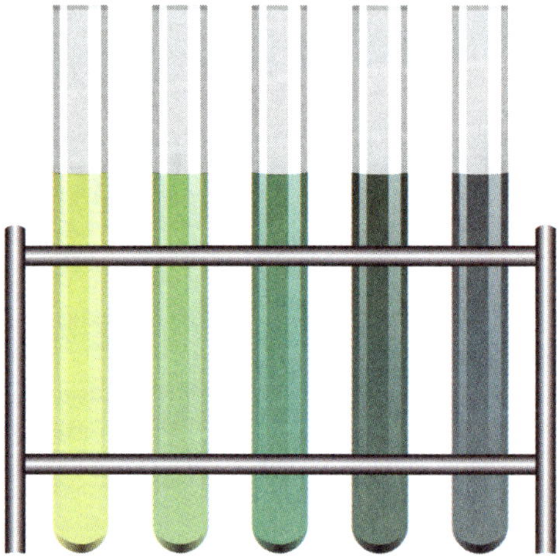

Figure 1 Concentration of a molecule.

Before measuring the amount of light absorbance of a sample, a researcher must first determine at what wavelength the spectrophotometer should be set. The wavelength setting depends on the color of solution under investigation. The wavelength setting is usually opposite on the color wheel of the visible light spectrum. For example, yellow solutions reflect yellow light but don't absorb it. Instead, they absorb other wavelengths of light. So, yellow solutions can be analyzed using a wavelength of 420 nm, the wavelength for violet light.

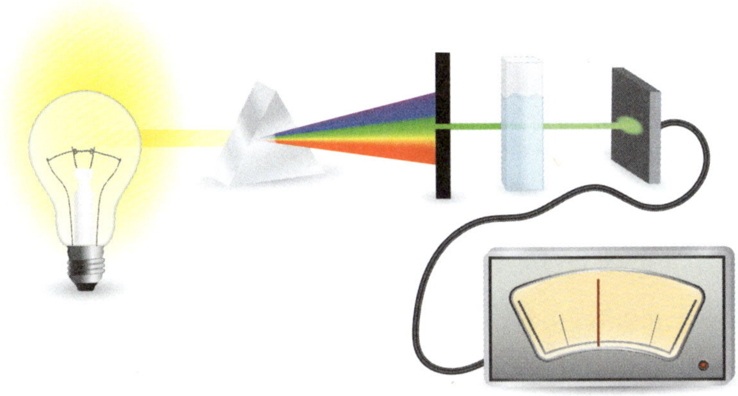

Figure 2 Light spectrum.

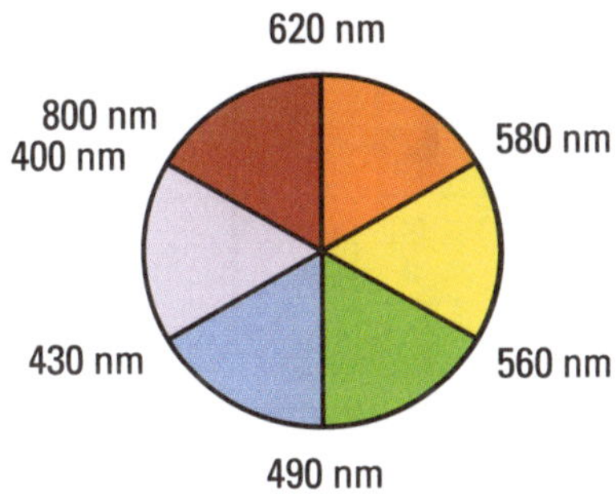

Figure 3 Color wheel.

During today's lab you will be constructing a standard curve by plotting known concentrations of methylene blue dye against their absorbance values obtained from the spectrophotometer. When completed, your standard curve should yield a nearly straight trendline with a high R^2 value (close to 1). You can then use the equation of this line to calculate concentrations of unknown solutions based on the measured absorbance value.

How to Use Micropipettes

Micropipettes enable you to measure small volumes of liquids. Using them to pipette accurately is a practiced skill that many students in the class are trying for the first time today. Don't hesitate to ask your TA for assistance if you haven't had much experience with micropipettes in previous classes. Wear gloves when doing this exercise.

Setting the Volume

1. Determine the volume range of your micropipette. What are the ranges of the micropipettes listed below?

   a. Volume range for 1,000 µL pipette ___200___ to ___1 000___

   b. Volume range for 200 µL pipette ___20___ to ___200___

   c. Volume range for 20 µL pipette ___2___ to ___20___

2. Turn the adjustment knob to set the desired volume, but **do not attempt to set it above the upper range or below the lower range!**

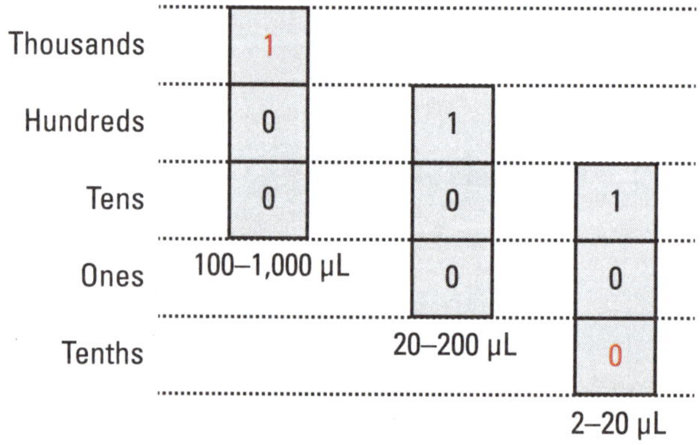

Figure 4 Setting volumes on micropipettes. According to the 100–1,000 µL micropipette window above, the micropipette is set to take up 1,000 µL of liquid. The 20–200 µL pipette is adjusted to transfer 100 µL of liquid, while the 2–20 µL micropipette is calibrated to take up 10 µL of liquid.

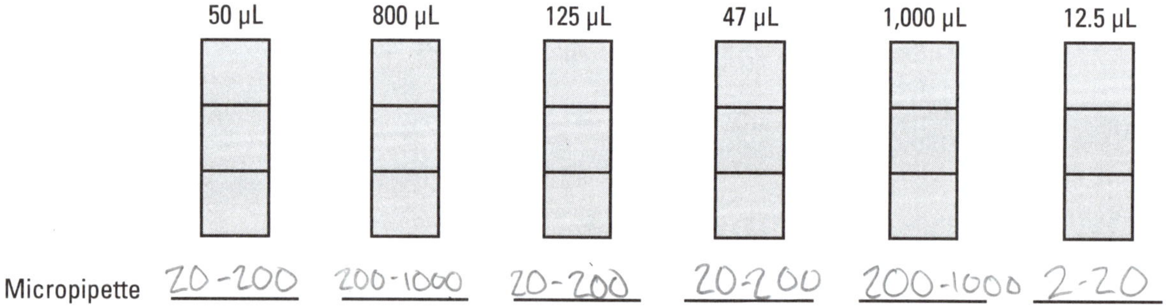

Figure 5 Adjusting volumes on micropipettes. Choose which micropipette you would use to measure the following volumes and write it in the blank space below each window. Fill in the figure with the correct numbers, as it would read in the micropipette window.

Taking Up (aspirating) Liquid

1. Attach an appropriately sized tip to your micropipette.
2. Press the plunger down to the **first stop.**
3. Immerse the pipette tip 2 mm into the liquid.
4. *Slowly* release the plunger to take up liquid.

Dispensing (ejecting) Liquid

Slowly press the plunger to dispense the liquid until you've reached the **second stop.**

Important Notes

- **Never** pipette liquid without attaching a tip to the micropipette.
- **Never** lay a micropipette down with liquid in the tip.
- Always keep the caps on bottles/tubes when not actively pipetting from them.
- Always keep the tip box closed when not actively getting a tip.

Micropipetting Practice Activity

1. Transfer 1,000 µL (1 mL) of colored water into a microcentrifuge tube.
2. Check the volume on the side of the tube. If you are above or below the 1.0-mL line, then you've made an error pipetting and should try again.
3. Repeat using 250 µL of colored water.
4. Each group member should take a turn practicing to make sure **all** group members feel comfortable using the micropipette.

Making Dilutions

Making dilutions is something that research scientists working in the lab do on a daily basis. It is an essential skill that takes practice at first. One way to make this task seem more approachable is to think about what you have, then think about what you want. In other words, what concentrations and volumes of a solution do you have initially, and what concentration and volumes of a solution do you want finally? This is written as the equations below:

Concentration(initial) × volume(initial) = concentration(final) × volume(final)

This equation is often abbreviated: $C_1 \times V_1 = C_2 \times V_2$

Let's say, for example, you have a stock solution of sugar dissolved in water with a concentration of 500 g/L.

1. How would you make 25 mL of a solution that is 375 g/L?

$C_1 \times V_1 = C_2 \times V_2$

$(500 \text{ g/L}) \times x = (375 \text{ g/L}) \times (25 \text{ mL})$

$x = 18.75 \text{ mL}$

18.75 mL stock solution needs to be mixed with 6.25 mL water

2. How would you make a solution that is ½ as concentrated (only 250 g/L) and has a final volume of 100 mL? (**Hint:** Use the formula $C_1V_1 = C_2V_2$.)

$500 \cdot x = 250 \cdot 100$

$\frac{500x}{500} = \frac{25000}{500}$

$x = 50$

3. How would you make a solution that has a final concentration of 125 g/L with a final volume of 80 mL? (**Hint:** Use the formula $C_1V_1 = C_2V_2$.)

$500 \cdot x = 125 \cdot 80$

$\frac{500x}{500} = \frac{10000}{500}$

$x = 20$

Standard Curves

Standard curves can help you find the concentrations of a substance in an unknown sample. The basic way this will work is to compare absorbance values of the unknown samples to absorbance values of samples with known concentrations.

Today, you'll be working with methylene blue dye. To get started making a standard curve, you need to have solutions with known concentrations of dye. You are provided with a stock solution with a concentration of 100 µg/mL methylene blue. Complete the table below to determine how to mix this stock dye with water to get the remaining known concentrations for your standard curve. (**Hint:** Use the formula $C_1V_1 = C_2V_2$.)

Table 1 Methylene blue standard curve.

Concentration of Methylene Blue (100 µg/mL stock)	Volume of Water (µL)	Volume of Methylene Blue Stock (µL)	Final Volume
0 µg/mL	2,000	0	2,000 µL
.5 µg/mL	1,990	10	2,000 µL
1 µg/mL	1,980	20	2,000 µL
2 µg/mL	1,960	40	2,000 µL
4 µg/mL	1,920	80	2,000 µL
8 µg/mL	1,840	160	2,000 µL

Work in teams of two to make the standard curve and measure absorbance values using the spectrophotometer. One member of each team should pipette the methylene blue dye while the other member will pipette the distilled water. Follow these steps:

1. Turn on the spectrophotometer using the power switch located on the back of the machine.

2. Using the appropriate micropipette, add the correct volume of distilled water and methylene blue dye to each test tube.

3. Gently swirl the tubes to mix the contents.

4. Once the spectrophotometer has warmed up, it should be set to absorbance (A) mode. If it is not, press the A/T/C button to select absorbance (A).

5. Change the wavelength to 650 nm by pressing the up "nm Δ" or down "nm ∇" buttons.
6. Wipe off your test tubes with a kimwipe to remove fingerprints.
7. Insert a "blank" test tube into the sample chamber and close the lid. Your "blank" should contain everything except what you want to measure. In this case, you want to measure dye in a solution of water, so water is your blank.
8. Press the "0 ABS / 100% T" button to set the blank to 0 absorbance.
9. Remove the blank from the chamber.
10. Insert your remaining samples and record the absorbance values in Table 2.

Table 2 Methylene blue absorbance values.

Concentration of Methylene Blue (µg/mL)	Absorbance 650 nm	
	Team A	Team B
0	0	0
.5	0.046	0.051
1	0.123	0.111
2	0.256	0.667
4	0.602	0.989
8	1.321	1.313

Clean Up

1. Dispose of solutions in the appropriate waste containers.
2. Wash tubes thoroughly with water and place the tubes upside down in the rack to dry.
3. Dispose of used tips, tubes, or kimwipes in the trash.
4. Wipe down benches with bleach.
5. Wash hands.

Plot absorbance on the *y*-axis vs. concentration on the *x*-axis for both teams below.

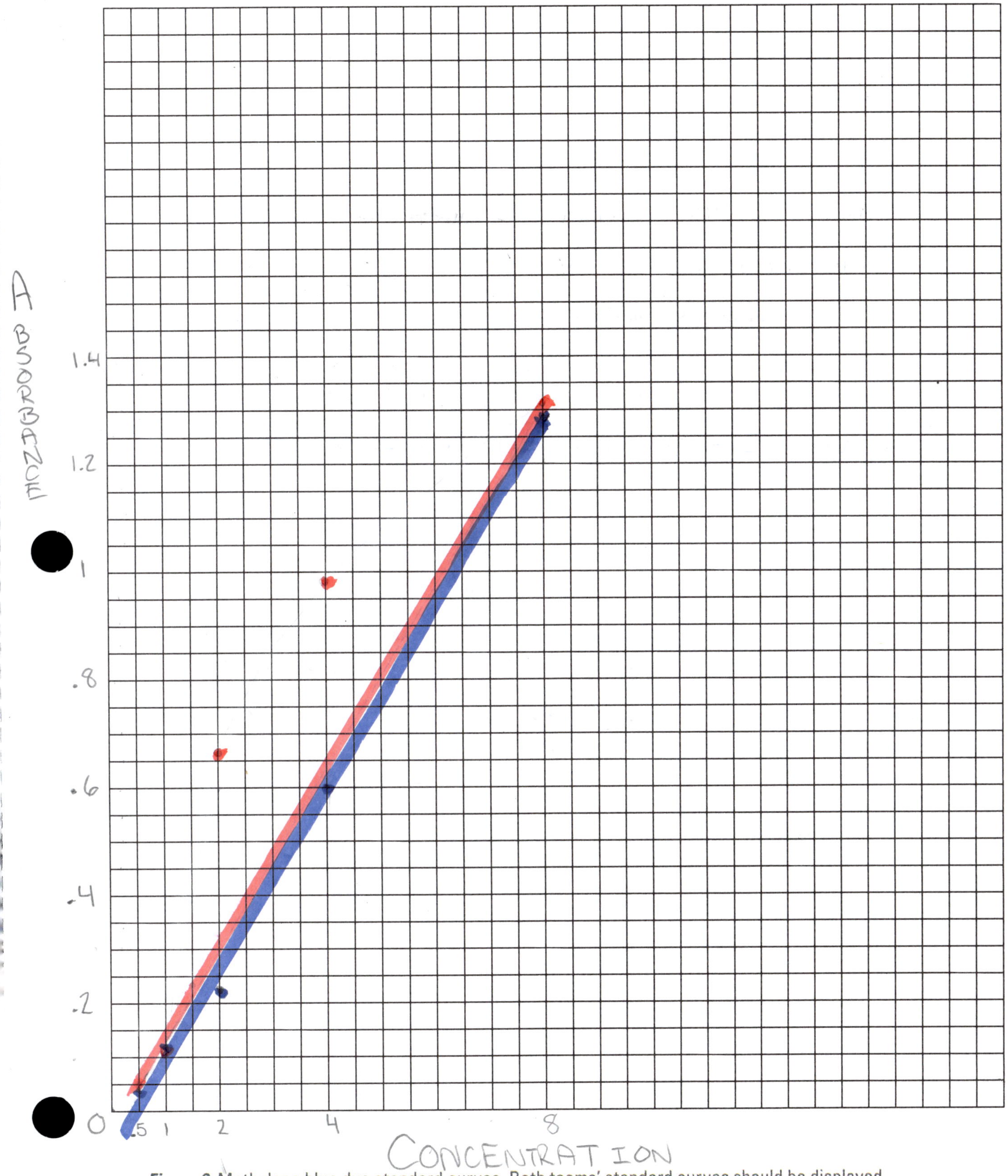

Figure 6 Methylene blue dye standard curves. Both teams' standard curves should be displayed.

Compare Curves

Compare the standard curves for Teams A and B and answer the following questions. This exercise will help you spot any errors you may have made and know how to prevent them in the future.

1. Did both teams get a similar standard curve? Why or why not?

   Yes, other than 2 and 4 being differant we had similar data

2. Were you able to draw a straight line through each team's data set? If not, what could you do differently to improve your standard curve?

   No, I feel like air bubbles or little errors in measuring

3. Were any of the data points outliers? How did you identify it as an outlier, and what could be the source of error resulting in this outlier?

   Yes, I have a couple outliers that aren't laying on the line that could be due to errors in measuring

Estimate Concentrations

Measure the absorbance values for samples that contain unknown concentrations of methylene blue and record them below. Estimate the concentrations using your hand drawn standard curve. Using your finger on the graph, find the unknown absorbance on the *y*-axis, trace a straight line horizontally to the standard curve, then trace a vertical line down to the *x*-axis. You've found the concentration! Record the concentrations below.

Table 3 Unknown sample calculated methylene blue concentrations.

Unknown Sample	Absorbance at 650 nm	Estimated Concentration of Methylene Blue Dye (µg/mL)
1	0.081	.6
2	0.338	2.4
3	0.991	4.1
4	0.571	3

Graph in Microsoft Excel

Now that you know what your standard curve should look like and how you use one to estimate concentrations, you can take it one step further. Instead of just estimating, you can calculate the concentration to get a more accurate value. Graphing the standard curve on a computer will help. Create a standard curve in Microsoft Excel.

Calculating Unknown Concentrations Using a Standard Curve

Standard curves can help you find the concentrations of a substance in unknown samples. Follow the example below to understand how this works. In the example below, researchers are determining the concentration of bovine serum albumin (BSA) in two different unknown samples. They generated a standard curve using known concentrations of BSA and used the same spectrophotometer settings to measure the absorbance of their unknowns. Standard "curves," like the one you just created with methylene blue and the one below, usually yield a straight line. The line is represented by the following equation:

$y = mx + b$

In the example below, the variables in the equation represent:

y = absorbance at 595 nm on the y-axis

m = slope of the line

x = concentration of BSA on the x-axis

b = the y-intercept

Using the equation of the line, it is possible to determine the concentrations of the unknowns by plugging in values for the known variables and solving for the missing variable, x or concentration.

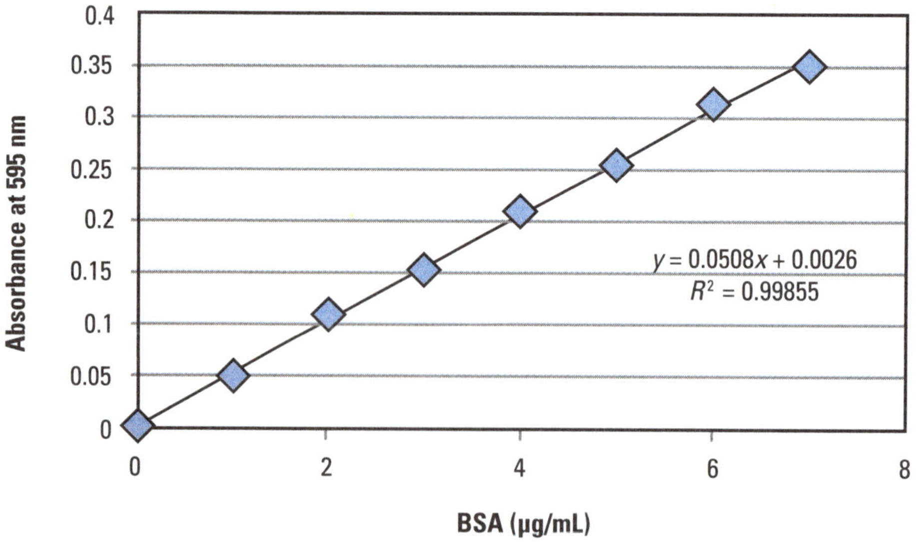

Figure 7 Bovine serum albumin (BSA) standard curve.

$\frac{2-1}{.256-.123} = \frac{1}{.133} = 7.52$

$y-1 = 7.52(x-.123)$ $y = 7.52x + .075$

Example 1

Unknown Sample 1 gave an absorbance reading of 0.200. Based on the standard curve above, what is the concentration of BSA in sample 1?

$0.200 = (0.0508)(X) + 0.0026$

$X = (0.200 - 0.0026)/0.0508$

$X = 3.886$

The concentration of BSA is 3.886 μg/mL.

Example 2

Unknown Sample 2 gave an absorbance reading of 0.133. What is the concentration of BSA in this sample?

$0.133 = (0.0508)(x) + 0.0026$

$(0.133 - 0.0026)/0.0508 = x$

$2.5669 = x$

Calculate Concentrations

Use your standard curve plotted in Excel to find the concentrations of unknown samples of methylene blue dye. Find the equation of the line, plug in knowns and measurements, and then solve for the appropriate variable to get the concentration.

Table 4 Unknown sample calculated methylene blue concentrations.

Unknown Sample	Absorbance at 650 nm	Calculated Concentration of Methylene Blue Dye (μg/mL)
1	0.081	.662
2	0.338	2.451
3	0.901	
4	0.571	

Assignments

- In-Class
  - Take good notes in your lab manual. Actively participate in lab.
- Homework
  - Complete and submit the Lab Skills homework assignment. Expectations for this assignment will be discussed in lab.
  - Read Lab 4: Investigating Enzymes—Mushroom Metabolism
  - Quiz

Lab 4 | Investigating Enzymes—Mushroom Metabolism

Assignments

In-Class

- ❒ Notes and participation

Homework

- ❒ Complete a Testing Hypotheses Worksheet based on the experiment you design and carry out. Submit the assignment using Turnitin on eCampus.
- ❒ Write Methods, Results, Figures, Discussion, and Introduction in scientific journal style. Include a Title section. Writing style is part of the grade. Use the rubric and information in **Appendix A** when writing.
- ❒ Read Lab 5: Scientific Literature
- ❒ Quizzes

Learning Goals

By the end of this lab, you will be able to:

- Present unique questions about the topic
- State complete scientific hypotheses for your questions
- Design experiments that are appropriate to test the hypotheses:
  - Incorporate proper constants and controls
  - Identify independent and dependent variables
- Carry out the experiment using proper lab techniques:
  - Pipette accurately
  - Correctly calculate chemical conversions and dilutions
  - Use standard curve
  - Collect and record data that will allow you to support or refute your hypothesis
- Communicate your research in scientific journal style
  - Accurately and completely describe the methods.
  - Present your results and figures
  - State whether or not your hypothesis was supported using the data you collect
  - Compose a discussion with logically drawn conclusions of your data and significance of your research
  - State whether or not your hypothesis was supported using the data you collect
  - Explain why your experiment is interesting or relevant by introducing background information

Introduction

Enzymes are proteins that function as catalysts, meaning they speed up biochemical reactions in living organisms. Some enzymes, like the ones found in digestive systems, break down large molecules into smaller ones. Others, like the ones that help build new DNA, use smaller molecules to construct larger ones. Without enzymes, essential processes in living things wouldn't happen, or they would occur very, very slowly. In fact, some reactions would take years without the appropriate enzyme, but when catalyzed, the reactions can occur in fractions of a second (Cooper 2000)! Simply put, without enzymes, organisms couldn't function.

Whether building or breaking apart, an enzyme's job is to convert substrates into products. The substrates bind to the active sites of an enzyme in such a way that they are held in the proper orientation for them to react. The resulting products that form are then released from the enzyme, leaving the enzyme ready to bind new substrates. Nearly every process that happens in living organisms involves enzymes, but each enzyme can only bind specific substrates that have complementary shapes, and thus each enzyme can only do a specific job. That's why, at any given moment, there are thousands of different enzymes at work inside cells (Cooper 2000)!

The activity of enzymes depends on several factors. Temperatures, pH, amount of enzyme, or other factors can affect the efficiency or rate of an enzymatic reaction. Enzymes are sensitive to the conditions around them. As the conditions change, so too will the rate of the reaction. While moderate changes might speed up a reaction, extremes could slow it down. For example, increasing the temperature a little might increase the rate of the reaction by increasing chances of substrate binding (due to increased kinetic energy of the molecules). However, increasing the temperature a lot could slow the reaction down—enzymes could denature and lose their shape that allows them to bind substrates.

Like all living organisms, mushrooms depend on enzymes for metabolism. One of these enzymes will be the focus of the lab today: Cellobiase. Cellobiase is an enzyme that breaks down the cellulose found in plant cell walls. Cellulose is a polysaccharide that is made up of hundreds to thousands of glucose molecules linked together. Fungi use cellobiase to break the cellulose into simple sugars, like glucose, that can be more easily metabolized and used as food.

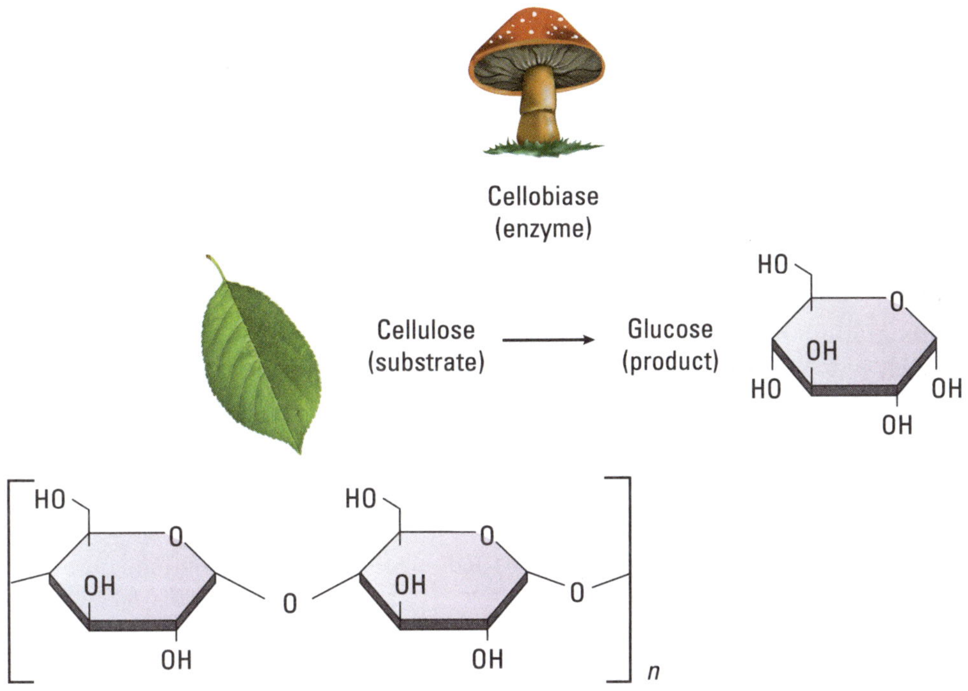

Figure 1

Mushrooms are made of a network of filaments called hyphae or mycelium. The mushrooms themselves are reproductive structures made of very dense hyphae, but most of the fungus is below ground. This is where the mycelium "mines" the environment for resources and food—including plant materials containing cellulose. As the mycelium grows outward into the soil, enzymes are released that begin to break down organic materials, like cellulose, into smaller molecules. These smaller molecules are absorbed by the hyphae and used within the fungus.

Mushroom species differ in where they live, how they get their energy, and what they use for "food." Most are saprophytes and live on dead, decaying organic matter from plants and trees. A few are parasitic—they grow on living trees and are harmful to them. Still others are mycorrhizal and form symbiotic associations with plants or trees; both the plant and fungus benefit by exchanging resources in these relationships. In the lab we have several species of mushrooms from grocery or specialty food stores. Below is a list of species that may be available in the lab today along with some information about each species, including what they may use for "food." (Your TA will update this list if there are any changes.)

Table 1

Cremini *(Agaricus bisporus)*	Grows in fields and on manure; saprophyte; (maturing mushroom)
Enoki *(Flammulina velutipes)*	Grows on stumps of dead hardwood trees (Chinese hackberry, ash, mulberry, persimmon); saprophyte
King Oyster *(Pleurotus eryngii)*	Grows on roots of herbaceous plants; saprophyte
Lion's Mane *(Hericium erinaceus)*	Grows on living hardwood trees (American beech, oaks); parasite or saprophyte
Oyster *(Pleurotus ostreatus)*	Grows on trunks of dead deciduous trees (beech, oaks); saprophyte
Portabella *(Agaricus bisporus)*	Grows in fields and on manure; saprophyte; (mature mushroom)
Shiitake *(Lentinula edodes)*	Grows on decaying deciduous trees (chestnut, oak, maple, beech, poplar, etc.); saprophyte
White Button *(Agaricus bisporus)*	Grows in fields and on manure; saprophyte; (immature mushroom)

Part 1: Questions and Hypotheses

You have the opportunity today to ask questions about the cellobiase enzymes contained in mushrooms. You might consider the fungus structure, species, cellobiase activity, or other factors in your question. Write out your question followed by a complete scientific hypothesis (if-and-then style). Given that a good hypothesis is testable, you'll need to know what's testable in our setting. Below are some hints that will help you complete this task.

Measuring Cellobiase

It is possible to measure cellobiase from mushroom tissues indirectly. You will not actually be able to measure the enzyme itself or the glucose products of the reaction when the typical substrate, cellulose, is used—these are colorless and not easily measured. However, there is another way to measure the activity of cellobiase. The substance p-nitrophenyl glucopyranoside can also bind to cellobiase as a substrate. When cellobiase catalyzes the reaction, this substrate breaks into its component products, glucose and p-nitrophenol. In a basic solution, p-nitrophenol is yellow and can be measured using the spectrophotometer; for every molecule of p-nitrophenol measured, one molecule of p-nitrophenyl glucopyranoside has been broken apart by cellobiase.

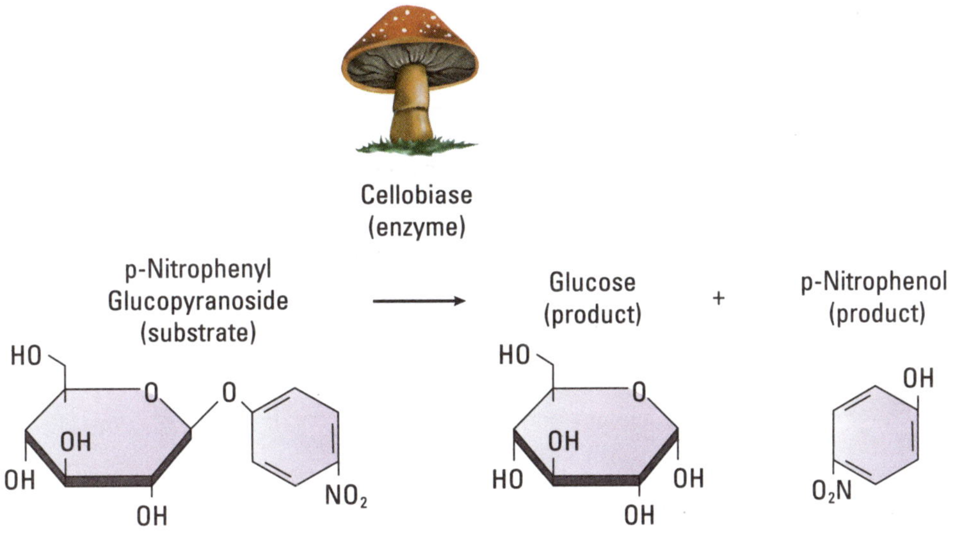

Figure 2

Recall from previous labs that standard curves generated using a spectrophotometer are used to quantify a substance. A spectrophotometer will give an absorbance value based on the intensity of color in solutions. You can quantify the cellobiase activity in mushroom samples, for example, by mixing mushroom extracts with substrate (p-nitrophenyl glucopyranoside), measuring absorbance values that correspond to the amount of product of the reactions (p-nitrophenol), and comparing those values to absorbance readings of solutions with known concentrations of the product. Details on how to set up cellobiase enzyme reactions and product standard curves are in tables below. Keep in mind that all groups will be expected to quantify activity of cellobiase enzymes.

Write your question and hypothesis below:

If mushrooms have a higher cellobiase activity in enviroments with dead wood, and we compare the p-nitrophenol concentrations of Cremini, King Oyster, and Oyster mushrooms at 25°C, then Oyster will have the most cellobiase activity compared to king oyster & cremini because it grows in the trunks of deciduous trees

If temperature and pH levels effect the activity of cellobiase enzymes in the shiitake mushroom and I compare the p-Nitrophenyl Glucopyranoside at two different temperatures then the activity of the enzyme should operate differantly

temperature- 45° AND 75°

Part 2: Designing the Experiment

Now that you have a general idea of how you will test your hypothesis, it is time to fill in the details. Below is information that will help you detail your methods.

Using Balances

The digital balances in lab are very sensitive and enable you to accurately weigh items. Be careful to treat them gently. Don't place heavy items like books on the balances. Here's how to use them:

1. Turn the balance on using the "On/Off button."
2. Place a weighing container onto the balance.
3. Press the Tare or "Zero" button.
4. When the balance reads "0.00" the material being weighed can be added to the weighing container.

Extracting Cellobiase from Mushroom Tissues

Follow these steps to obtain a mushroom extract that contains cellobiase.

1. Weigh out a 0.5-g tissue sample of each mushroom you wish to test.
2. Transfer the sample to a mortar and add 1 mL of extraction buffer. Using a pestle, grind the mushroom sample with the buffer until it forms a slurry or paste.
3. Transfer the slurry or paste to a 1.5-mL microcentrifuge tube. Centrifuge the sample for 3 minutes at 13,000 rpm.

Setting up Cellobiase Reactions

1. For each reaction, mix 62.5 µL of the supernatant from the mushroom extract (containing cellobiase enzyme) with 750 µL of 1.5 mM p-nitrophenyl glucopyranoside (substrate) solution in a 1.5-mL microcentrifuge tube. Gently invert the tube several times to mix the reaction. Let the tube incubate for the amount of time chosen for your experiment.
   - **Note:** Students interested in studying a single reaction over time will need a larger volume of reaction mix (substrate + enzyme). Mix 250 µL of the supernatant from the mushroom extract (containing the cellobiase enzyme) with 3 mL of 1.5 mM p-nitrophenyl glucopyranoside (substrate) solution in a borosilicate glass test tube.

2. Prepare a "blank." Mix 62.5 μL the supernatant from the mushroom extract (containing cellobiase enzyme) with 750 μL of 0 mM substrate solution (containing ***no*** p-nitrophenyl glucopyranoside) in a microcentrifuge tube.

3. While incubating the reactions, pipette 500 μL of stop solution into labeled cuvettes.

   ■ **Note:** If you are using short timepoints, pipette the stop solution into the cuvettes prior to setting up the reaction.

4. After the desired incubation time, transfer 500 μL of the enzyme reaction mix into the cuvettes with the stop solution.

5. Follow the instructions for "Quantifying p-Nitrophenol" below to set up the spectrophotometer and record absorbance values. Compare absorbance values for your samples to the standard curve to determine p-nitrophenol concentration as an indication of cellobiase activity.

Table 2 200x = 500x

	Sample	Volume of Reaction Mix (Mushroom Extract + Substrate) (μL)	Volume of Stop Solution (μL)	Final Volume in Cuvette (μL)	Absorbance at 410 nm
Cremini	1	500 μL	500 μL	1,000 μL	.825
Cremini-B	2	500 μL	500 μL	1,000 μL	0
Oyster	3	500 μL	500 μL	1,000 μL	.826
Oyster-B	4	500 μL	500 μL	1,000 μL	0
King Oyster	5	500 μL	500 μL	1,000 μL	.742
King Oyster -B	6	500 μL	500 μL	1,000 μL	0
	7	500 μL	500 μL	1,000 μL	

Controlling Cellobiase Reaction Times

Enzymatic reactions are often time sensitive. Typically, the more time the reaction has to run, the more product you'll get. Depending on your question, you may want to measure all reactions after they run the same amount of time (time would be a constant), or you

375 75 150 300

may want to measure at multiple time points (if you are interested in comparing rates or enzymatic activity over time). Cellobiase reactions will produce measureable products within minutes. As you design your experiment, pick a reaction time around 15 minutes. Students interested in multiple time points should plan to measure in increments starting before and going beyond 15 minutes. You will stop the cellobiase reactions by adding the reaction mix (extracts containing enzyme + substrate) to stop solution. The stop solution is basic (high pH) and denatures the cellobiase enzymes. The high pH also causes p-nitrophenol to turn yellow.

Setting up Standards

Use the information in the following table to mix standards for constructing a standard curve. A stock solution of p-nitrophenol will be provided at a concentration of 200 nM. Use the $C_1V_1 = C_2V_2$ calculation to determine how to dilute the stock solution with water to get the remaining known concentrations of standards. (**Hint:** V_2 = the volume of p-nitrophenol + volume of water = 500 µL.)

1. Make dilutions of p-nitrophenol according to your calculations in the table below. Mix the stock with water for each standard solution directly in labeled cuvettes.
2. Add 500 µL of stop solution to each cuvette.
3. Follow the instructions for "Quantifying p-Nitrophenol" to set up the spectrophotometer and record absorbance values.

500x = 500y

Table 3 Standards for calculating p-nitrophenol concentration.

200x = 500y

p-Nitrophenol Concentration (nM)	Volume of 200 nM p-Nitrophenol (µL)	Volume Water (µL)	Stop Solution (µL)	Final Volume in Cuvette (µL)	Absorbance at 410 nm
0	0 µL	500 µL	500 µL	1,000 µL	0
25 nM	62.5	437.5	500 µL	1,000 µL	.149
50 nM	125	375	500 µL	1,000 µL	.180
100 nM	250	250	500 µL	1,000 µL	.388
200 nM (provided stock solution)	500	0	500 µL	1,000 µL	.685

Quantifying p-Nitrophenol

To determine the concentration of p-nitrophenol as an indication of cellobiase activity, you'll need to create standard curves using absorbance values from a spectrophotometer. Use these general instructions for your standard solutions and your samples.

1. Prepare mushroom extracts and set up cellobiase reactions. (See above.)

2. Set up standards for the standard curve. (See above.)

3. Turn on the spectrophotometer using the power switch located on the back of the machine. Once the machine has warmed up, it should be set to absorbance (A) mode. If it is not, press the A/T/C button to select absorbance (A).

4. Set the spectrophotometer to the appropriate wavelength using the "nm Δ" or down "nm ∇" buttons. For p-nitrophenol, read at 410 nm.

5. Insert a "blank" test tube into the sample chamber and close the lid. Press the "0 ABS/ 100% T" button to "blank" the machine. Remove the blank test tube from the chamber.

   - The "blank" tube contains everything except the thing you are trying to measure. If you are measuring p-nitrophenol, your blank will have solution with no p-nitrophenol, but have everything else. In this experiment, you will probably need two blanks—one when you are measuring p-nitrophenol with your standard curve, and another when you are measuring p-nitrophenol in your mushroom extracts.

6. Place the cuvettes in the spectrophotometer one at a time and record absorbance values for each one at 410 nm.

   - Don't blank the spectrophotometer between samples—this is unnecessary.
   - Do blank the spectrophotometer when your background solution has changed or if you are reading different wavelengths.

7. Plot a standard curve using the known concentrations of your p-nitrophenol standards vs. their corresponding measured absorbance values. Find the equation of the line that represents the standard curve.

8. Compare your mushroom samples to the standard curve to determine p-nitrophenol concentration as an indication of cellobiase activity. Use the equation of the line from the standard curve to calculate the concentrations.

Write out your plans for your experiment below:

Collect a .5g sample of shiitake tissue →  Add 1mL of extraction buffer

↓

Transfer into 1.5 mL tubes then centrifuge for 3 minutes 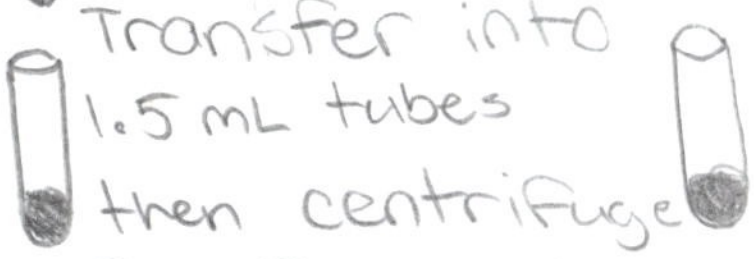

←

Mix 62.5 μL of the supernatant with 750 μL of substrate

↓

Insert tube

10 mins of incubation

→

prepare a blank

mix 62.5 μL of supernatant with 750 μL of substrate

↓

Add 500 μL of the stop solution

↓

Add 500mL of enzyme mix

↓

Set up standard

METHOD

Collect .5 grams of your 3 mushrooms. Add 1mL of extraction buffer to the sample. Transfer into 1.5 mL tubes. Centrifuge for 3 minutes. Mix 62.5 μL of the supernatant with 750 μL of substrate. 10 mins of incubation. Prepare a blank. Mix 62.5 μL of supernatant with 750 μL of substrate. Add 500 μL of the stop solution. Add 500mL of enzyme mix. Set up standard.

= +x

-x =

Part 3: Carry out the Experiment

Once your plans have been approved by your TA, you may begin your experiment. Be sure to use proper personal protective equipment (lab attire, gloves, eye protection, etc.). Also know that you are responsible for keeping the lab neat and tidy—a clean lab is a safe lab!

Use the space below to take notes and record data:

Clean Up

1. Pour the contents of test tubes in the appropriate waste containers. Do not pour it down the drain!
2. Thoroughly rinse all labware for use by the next class.
3. Wipe off lab benches with 10% bleach and wash your hands!

Part 4: Graphing and Statistics

An important part of any scientific study is analyzing and presenting data. In lab you will learn how to use statistics to make important comparisons and conclusions. You will also learn how to use Microsoft Excel to make graphs and figures.

Assignments

- In-Class
  - Take good notes in your lab manual. Actively participate in lab.
- Homework
  - Complete a Testing Hypotheses Worksheet using notes you took in lab. Type it using the electronic form for the assignment. Submit through the Turnitin link on eCampus. This is an independent homework assignment.
  - Present Methods, Results with Figures, Discussion, and Introduction in scientific journal style according to the rubric. Include a Title section. Writing Style is also part of the grade. Use the rubric and guides in **Appendix A** when writing these sections. It is an independent homework assignment; students are expected to write on their own. Students are expected to create their own figures in Excel—no sharing among group members or other students. This is an independent homework assignment.
    - Sections of a complete lab report written in scientific journal style will be assigned throughout the semester. See the schedule of assignments on eCampus and listen in lab for due dates. Expectations for these assignments will be discussed in lab, so please pay attention and take notes!
  - Read Lab 5: Scientific Literature
  - Quizzes

References

Cooper GM. 2000. The cell: a molecular approach. 2nd ed. Sunderland (MA): Sinauer Associates.

Lab 5 | Scientific Literature

Assignments

In-Class

- ❒ Notes and participation

Homework

- ❒ Incorporate information from scientific literature into your lab report.
- ❒ Read Lab 6: Bacterial Transformation
- ❒ Quiz

Learning Goals

By the end of this lab, you will be able to:

- Distinguish primary from secondary literature
- Recognize paraphrases from scientific articles
- Use CSE citation formats for references
- Find scientific articles using databases
- Identify questions, hypotheses, and experimental elements in articles
- Use and cite scientific literature appropriately when writing in scientific journal style

Introduction

As a biology student, you will search the scientific literature to find information. As you read, you'll notice that science can be communicated in a variety of forms that differ based on their intended purpose and audience. When it comes to published sources, there are two broad categories of scientific literature: primary and secondary.

Primary literature is usually in the form of scientific articles published in peer-reviewed journals. These articles usually have a format that includes Introduction, Methods, Results, Discussion, and Reference sections. Scientists use these articles to communicate with other scientists. They are probably the most widely used sources of information for scientists to learn more about current research in their field, to design next-step experiments, and to reference previous work in their own publications. Given that the primary literature is usually directed toward a scientific audience, it assumes an understanding of terms or language ("jargon") and scientific concepts that are specific to the field in which it is written.

Primary research articles present novel data collected from experiments and studies conducted by the author(s) themselves. Rarely research articles will have a single author, but much more frequently, you'll see a list of several authors on a paper. Biology is a collaborative science, and there is usually a team of scientists that contributed to the research presented. Often there are scientists from multiple universities or research institutions working together. You will see the authors' affiliations indicated with symbols after their names.

One of the important features of primary scientific research articles is that they are **peer-reviewed.** During the peer review process, scientists submit a manuscript to a journal, and then the editor asks reviewers to comment on it. The reviewers are usually scientists in the field and have the background knowledge to spot flaws in experiments, techniques, assumptions, or other issues. They then will recommend to the editor whether the paper be accepted, rejected, or accepted under the condition that the original authors make revisions to the manuscript. Sometimes the revisions might be simple and involve editing the writing; other times the revisions may require collecting additional data to support conclusions. The editor will then take the reviewers comments and recommendations into consideration and make the final decision about whether or not to publish the article. This type of rigorous screening of scientific literature ensures that work published in scientific journals is of high quality. Only work that is acceptable by experts in the field should be published and shared with others.

Just like scientists writing a research article, you will need to use primary literature when you are writing lab reports; in the Introduction you will frame your experiments in the context of others' published work, and in the Discussion you'll interpret your findings and compare to what's already known. The majority of references in your lab reports will be sources found in the primary literature. However, primary sources aren't the only type of scientific literature you'll come across.

Common **secondary sources** for scientific information are textbooks, magazines, websites, and review articles. Secondary sources are based on primary sources, but have a much different target audience and purpose. These sources are usually written for both scientists and non-scientists alike, including general audiences with little or limited knowledge about the topic presented. They are intended to inform about a topic, but usually have less detail than a primary article. For example, you won't find the specifics of how a particular experiment was carried out in a secondary article, though you might know what treatment groups were compared and what was measured. In addition to being less detailed, secondary articles and sources are often easier to read as they tend to contain less jargon and aren't as formal in style.

Some secondary sources summarize and interpret information from several primary sources. That makes them good places to start reading for students just beginning to learn about a topic. **Review articles** are a good example of secondary literature, as they take data from

-The type of disorder & info on that disorder
- Patients medical info
- The effects of therapeutic genes

multiple primary sources and combine it into one document. They can also be a good place to find primary articles since they will include a reference list of the primary literature cited in the article. While review articles are found in peer-reviewed journals and are reliable sources of information, they are not considered primary sources since the authors did not conduct the research themselves.

Caution should be used when examining secondary literature, as it often contains interpretations about someone else's data or could discuss results out of context. Therefore, secondary articles can sometimes state invalid conclusions about the work they reference or may even fail to correctly cite studies. This is why it is important to go back and read the primary literature that the secondary article references and draw your own conclusions. For lab reports, if you choose to cite any secondary sources, stick to more reliable sources, like textbooks or review articles. Steer clear of citing websites, sources that may have potential motives to persuade you, or sources with uncertain reliability.

Comparing Primary and Secondary Articles

It is important that you are able to recognize features of primary and secondary articles and can distinguish them. Your TA will provide you with articles. In the table below write defining characteristics of primary and secondary articles. Based on the information you listed in your table, determine which of your articles is a primary source and which is a secondary source.

Table 1

Elements	Primary (A)	Secondary (B)
Purpose and Audience	Information for other scientists	General population to inform
Format, Writing Style, and Language	Lab Report Format	Broad, easy to understand
Authors	Multiple scientists doing the research	Journalist
Type of Publication	peer re-viewed Scientific Research Article	Review Article / Website

PAGE 105 IN-TEXT CITATION

Using and Referencing Sources

When scientists are writing papers or making presentations, they often reference work published by others. You'll be expected to do the same as you prepare lab reports and some other assignments. Properly referencing others' work and publications is essential. If you don't reference, it is assumed that the work and ideas you present are your own. That means if you have taken phrases or ideas from sources and put them in your own document without citing them, you are stealing intellectual property. This is plagiarism. It is a serious offense with consequences that range from grade reduction on an assignment to expulsion from the University (see WVU's Student Conduct Code on Academic Dishonesty). These consequences may sound scary, but there are two things you can do to make sure that you avoid any such situation. First, paraphrase from sources, and second, cite properly.

Paraphrasing—What Do *You* Have to Say?

Paraphrasing means to put ideas into your own words. You will likely find yourself reading journal articles or books and want to present information you find as it relates to your topic. You'll likely do this in the introduction of lab reports to provide rationale for your experiment and in the discussion section to compare your findings with others'. It may be tempting to use the exact phrases found in the sources, but don't. Instead, as you comb through sources, think carefully about the information you've read. Be sure you understand it, then synthesize the information and organize your own thoughts. Then present what **you** want to say about it in your own words; this is how you paraphrase. You should never take text verbatim without citation, but you should also avoid using quotations—even if you cite the original source. While quotes may be acceptable in some classes, for this class you are expected to write in your own words.

When you are paraphrasing, be sure you are going beyond just rearranging phrases or substituting words. Take a look at the example below to compare appropriately paraphrased and poorly paraphrased (plagiarized) sentences.

- Original passage from Antoni et al. (2007):

  Yeast fermentation of glucose syrups to ethanol has been well progressed in recent years. Inhibitor sensitivity, product tolerance, ethanol yield and specific ethanol productivity have been improved in modern industrial strains to the degree that up to 20% (v/v) of ethanol are produced in present-day industrial yeast fermentation vessels from starch-derived glucose.

- Student sentences referencing the original source:
    - *Up to 20% (v/v) of ethanol are produced in present-day industrial yeast fermentation vessels from starch-derived glucose due to inhibitor sensitivity, product tolerance, ethanol yield and specific ethanol productivity that have been improved in modern industrial strains (Antoni et al., 2007).*

      This sentence is plagiarized; the order of the last sentence was changed and very little wording was modified.

    - *Yeast fermentation of glucose syrups to ethanol has been well advanced in recent years (Antoni et al., 2007).*

      This sentence is plagiarized; only one word was changed.

    - *Several improvements have been made to industrial strains of yeast that enable them to withstand high ethanol concentrations (up to 20% (v/v)) in fermentation vessels (Antoni et al., 2007).*

      This sentence is paraphrased; the original paper is cited as the source of ideas, but wording is the student's.

Practice with "Paraphrases"

Read the original passages, and then identify which of the statements that follow are appropriately paraphrased. Describe the problems with statements that are not appropriately paraphrased.

1. Original passage from Little et al. (2012):

   Traditionally, subterranean termites (Isoptera: Rhinotermitidae: *Reticulitermes*) have been considered to be important in forest succession and nutrient cycles in the southeastern US nearly exclusively from a decomposition standpoint. Little is known about the role of subterranean termites in relation to forest health and succession. In particular, no investigations of the role of subterranean termites during the early stages of tree death are present in the literature. The possibility that subterranean termites interact with bark beetles, for instance, remains largely unexplored despite the economic and ecological importance of both organisms.

   a. Despite the economic and ecological importance of both organisms, the possibility that subterranean termites interact with bark beetles, for instance, remains largely unexplored (Little et al. 2012).

   b. Subterranean termites and bark beetles are both important in regards to forest succession and health but it is not known if they interact with each other (Little et al. 2012).

c. The likelihood that subterranean termites interact with bark beetles remains for the most part unexplored despite the economic and ecological importance of both of these organisms (Little et al. 2012).

2. Original passage from Fernandez-Rodriguez et al. (2017):

In nature, single cells can contain dozens of light sensors that respond to the rainbow of colors as well as near-UV and far-infrared light, and these have been applied to eukaryotes and animals. Individually, these light sensors have been used to remotely control pathways in a bioreactor, induce lysis, and implement genome editing, and also have biomedical applications.

a. Light sensors can be engineered to respond to the full spectrum of light for use clinically and in biomedical research, such as controlling pathways of cellular response (Fernandez-Rodriguez et al. 2017).

b. Single eukaryotic cells in nature respond to the rainbow of colors, in addition to near-UV and infrared light. These light sensors individually have been used to control pathways remotely to induce lysis, edit genomes, control bioreactor pathways, and have applications in the biomedical field (Fernandez-Rodriguez et al. 2017).

c. Cells can contain dozens of light sensors that react to the colors of the rainbow, UV, and infrared light. This has been applied in eukaryotes, including animals. These light sensors have been used to regulate cell processes to induce cytolysis, edit DNA, influence bioreactor pathways, and have clinical medicine applications.

Citing Sources—Give Credit Where Credit Is Due

Even when you are paraphrasing and using your own words, you must cite the original source. For this course, you will format citations using the Council of Science Editors (CSE) name-year system. This CSE format or very similar styles are widely used among scientific journals that publish on various biology topics. You will use CSE format for in-text citations and for a Literature Cited or References section at the end of your lab reports. The references listed at the end of the document are unnumbered and are listed alphabetically by author.

Practice with CSE Citation Formats

When writing lab reports in the style of a scientific journal, you'll most often be citing primary literature. Answer the following questions to practice with this format. See **Appendix B:** CSE Council of Science Editors (CSE) Name-Year Style for Formatting References for examples of formatting.

1. Which of the following citations is correctly formatted? Look carefully! Make sure all information is present and punctuation is correct.

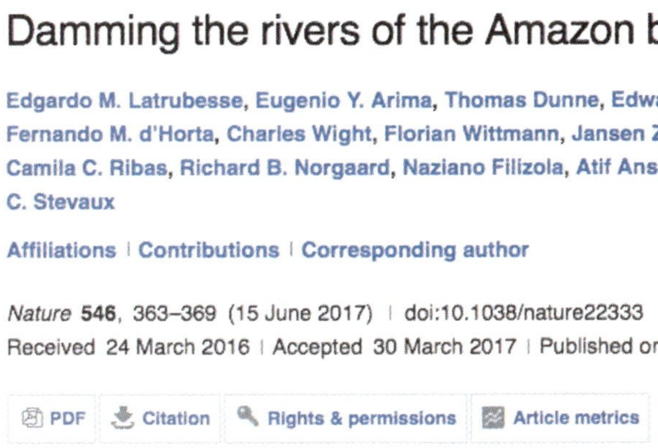
Damming the rivers of the Amazon basin

Edgardo M. Latrubesse, Eugenio Y. Arima, Thomas Dunne, Edward Park, Victor R. Baker, Fernando M. d'Horta, Charles Wight, Florian Wittmann, Jansen Zuanon, Paul A. Baker, Camila C. Ribas, Richard B. Norgaard, Naziano Filizola, Atif Ansar, Bent Flyvbjerg & Jose C. Stevaux

Affiliations | Contributions | Corresponding author

Nature **546**, 363–369 (15 June 2017) | doi:10.1038/nature22333
Received 24 March 2016 | Accepted 30 March 2017 | Published online 14 June 2017

PDF | Citation | Rights & permissions | Article metrics

Figure 1

Latrubesse EM, Arima EY, Dunne T, Park E, Baker VR, d'Horta FM, Wight C, Wittmann F, Zuanon J, Baker PA, et al. 2017. Damming the rivers of the Amazon basin. Nature. 546:363–369.

Latrubesse EM, Arima EY, Dunne T, Park E, Baker VR, d'Horta FM, Wight C, Wittmann F, Zuanon J, Baker PA, et al. Damming the rivers of the Amazon basin. 2016. Nature. 546:363–369.

Latrubesse EM, Arima EY, Dunne T, Park E, Baker VR, d'Horta FM, Wight C, Wittmann F, Zuanon J, Baker PA, Ribas CC, Norgaard RB, Filizola N, Asar A, Flyvbjerg B, Stevaux JC. Damming the rivers of the Amazon basin [Internet]. Nature. [Cited: 21 Jun 2017] Available from: http://www.nature.com/nature/journal/v546/n7658/full/nature22333.html DOI:10.1038/nature22333

2. Look at the image of the article below. Find the information needed for citations and write in-text and end citations out in the appropriate CSE format.

Clinical Nutrition 36 (2017) 1097–1104

Contents lists available at ScienceDirect

Clinical Nutrition

ELSEVIER

journal homepage: http://www.elsevier.com/locate/clnu

CLINICAL NUTRITION

Original article

Associations between micronutrient intakes and gut microbiota in a group of adults with cystic fibrosis

CrossMark

Li Li [a], Lutz Krause [b], Shawn Somerset [a, c, *]

[a] School of Medicine, Griffith University, Brisbane, Queensland, Australia
[b] The University of Queensland, Diamantina Institute, Translational Research Institute, Woolloongabba 4102, Brisbane, Queensland, Australia
[c] School of Allied Health, Faculty of Health Sciences, Australian Catholic University, Brisbane, Queensland, Australia

ARTICLE INFO

Article history:
Received 7 December 2015
Accepted 30 June 2016

Keywords:
Cystic fibrosis
Gut microbiota
Antioxidant vitamins
Inflammation

SUMMARY

Background: Cystic fibrosis (CF) involves chronic inflammation and oxidative stress affecting mainly the respiratory and digestive systems. Survival rates for CF have improved with advances in treatment including nutritional interventions such as micronutrient supplementation. Diet can modulate gut microbiota in the general population with consequences on local and systemic immunity, and inflammation. The gut microbiota appears disrupted and may associate with pulmonary status in CF. This study investigated associations between micronutrient intakes and gut microbiota variations in a group of adults with CF.
Methods: Faecal microbiota of sixteen free-living adults with CF was profiled by 16ss rDNA sequencing on the GS-FLX platform. Associations were tested between UniFrac distances of faecal microbiota and time-corresponding micronutrient intakes. Associations between relative abundances of bacterial taxa and micronutrient intakes (those showing significant associations with UniFrac distances) were examined by Spearman correlation.
Results: Unweighted UniFrac distances were associated with intakes of potassium and antioxidant vitamins C, E and beta-carotene equivalents, whereas weighted UniFrac distances were associated with antioxidant vitamins riboflavin, niacin equivalents, beta-carotene equivalents and vitamin A equivalents. Intakes of beta-carotene equivalents, vitamin C, vitamin E, niacin equivalents and riboflavin correlated negatively with *Bacteroides* and/or its corresponding higher level taxa. Intakes of beta-carotene equivalents and vitamin E also positively correlated with *Firmicutes* and specific taxa belonging to *Firmicutes*.
Conclusion: Some micronutrients, particularly antioxidant vitamins, correlated with gut microbiota variations in the studied cohort. Further research is required to clarify whether antioxidant vitamin intakes can influence CF gut microbiota and potential clinical/therapeutic implications in CF.

Figure 2 (Li et al. 2013)

- In-text: Survival rates for CF have improved with advances
- End reference: Li L, Lutz K, Shawn S, et al. 2017. Associations between micronutrient intakes and gut microbiota in a group of adults with cystic Fibrosis. Clinical Nutrition. 36:1097-1104

Searching Databases and Websites for Scientific Articles

For upcoming assignments, you will need to research information found in scientific articles. This exercise will give you practice finding and reading primary articles.

1. Choose a biological topic of interest to you and your group. Write it below, then list keywords that will help you narrow your search on your specific topic.

Glycosidases in The American Oyster Digestive Tract

2. Find a peer-reviewed, primary research article on your topic. Use academic databases like Google Scholar, Academic Search Complete, Web of Science, or the online search features of the WVU Libraries (*www.libraries.wvu.edu*) to find articles. Write the in-text and end reference citations for your article using CSE format.

   Note: You should focus on articles where the authors have conducted the research themselves. For this assignment, be sure you do not select a review article or other secondary source. If you are unsure if you have a primary article, ask your TA.

   - In-text:
   - End reference:

Assignments

- In-Class
  - Take good notes in your lab manual. Actively participate in lab.
- Homework
  - Incorporate information from scientific literature into your lab report. Expectations will be discussed in lab, so pay close attention and take good notes!
  - Read Lab 6: Bacterial Transformation
  - Quiz

References

Antoni D, Zverlov VV, Schwarz WH. 2007. Biofuels from microbes. Appl Microbiol Biotechnol. 77:23–35.

Fernandez-Rodriguez J, Moser F, Song M, Voigt C. 2017. Engineering RGB color vision into Escherichia coli. Nat Chem Biol. 13:706–708.

Little NS, Riggins JJ, Schultz TP, Londo AJ, Ulyshen MD. 2012. Feeding preference of native subterranean termites (Isoptera: Rhinotermitidae: Reticulitermes) for wood containing bark beetle pheromones and blue-stain fungi. J Insect Behav. 25:197.

Lab 6 | Bacterial Transformation

Assignments

In-Class

- ❐ Notes and participation

Homework

- ❐ Answer the Transformation Questions. Submit the assignment using Turnitin on eCampus.
- ❐ Complete a Testing Hypotheses Worksheet based on the experiment you design and carry out. Submit the assignment using Turnitin on eCampus.
- ❐ Prepare a presentation according to your TA's instructions.
- ❐ Read Lab 7: Science of Opioid Dependence
- ❐ Quizzes

Learning Goals

By the end of this lab, you will be able to:

- Present unique questions about the topic
- State complete scientific hypotheses for your questions
- Design experiments that are appropriate to test the hypotheses:
  - Incorporate proper constants and controls
  - Identify independent and dependent variables
- Research protocols for bacterial transformation:
  - Understand the reasons for performing steps
- Carry out the experiment using proper lab techniques:
  - Use aseptic technique for pouring agar plates and handling bacterial cultures
  - Collect and record data that will allow you to support or refute your hypothesis:
    - Calculate transformation efficiency
- State whether or not your hypothesis was supported using the data you collect
- Deliver a presentation based on your experiment and results as well as research from the scientific literature

Introduction

A bacterial transformation is the process of altering the genetic makeup of bacteria by introducing foreign DNA. In laboratory settings, bacterial transformations involve the transfer of plasmid DNA into the target organism. This process results in a genetically modified organism (GMO), also called a transformant.

A plasmid is a small, circular DNA molecule that is not part of a chromosome (i.e., plasmid DNA is not covalently bound to chromosomes) and replicates independently from chromosomes. Plasmids may contain genes that are beneficial to the survival of its host, such as genes that code for antibiotic resistance. When bacteria take up plasmids, the genes encoded on the plasmid will be transcribed and translated into functional proteins.

A common plasmid used in biotechnology is the pGLO plasmid, which contains genes that code for green fluorescent protein (GFP) and β-lactamase. The GFP and β-lactamase genes are not naturally found in *Escherichia coli,* a species of pathogenic bacteria. When the pGLO plasmid is incorporated into *E. coli* cells, the bacteria will transcribe both genes (GFP and β-lactamase) into mRNA and translate them into functional proteins. GFP glows bright green when exposed to UV light, while β-lactamase detoxifies the antibiotic ampicillin. Bacterial cells that have been transformed with the pGLO plasmid and express these genes will not only glow, but can grow in the presence of ampicillin. Because of this, researchers have the ability to selectively culture only transformed bacteria.

Part 1: Researching Bacterial Transformation Protocols

This lab involves setting up bacterial transformations. You will design and conduct an experiment to transform *E. coli* with the pGLO plasmid—you will make a GMO! Before you get started, it is important that you know the basic steps of a bacterial transformation protocol and the importance of each step. Search the Internet or other sources to answer the following questions.

1. What is the overall goal in conducting a bacterial transformation?

It is an important step in the cloning of the molecule, the main goal is to produce and multiply molecules of DNA

2. Describe why each of the following is important in a bacterial transformation.

   a. DNA - All transformations are influenced by the concentration of DNA

   b. Chemically competent *E. coli* cells

   c. Heat shock - conventional protocols bacteria transformation require a brief heat shock for maximum uptake of plasmid DNA.

   d. Medium (plural: media)
   Useful for the bacteria to go

   e. Selectable marker (i.e., ampicillin resistance)
   Tests the bacterias resistance

3. Using reliable sources, briefly describe the process of how bacteria are transformed in the laboratory. In five sentences or fewer, state a general method.

4. What is transformation efficiency? How is it calculated?

5. Find three sources detailing a protocol for a bacterial transformation. Compare those protocols. How do the protocols differ?

All Heat shock & temperatures are similar
Differant levels of SOC are piped into bacteria but at the same temp of 37°C.
The starter is shaken between 200-300 rpm
Incubation is done 10-15 min
Add cells to sterile tubes.

6. List several variables that may affect the efficiency of a bacterial transformation. What steps could you change when modifying the protocol? How would you expect transformation efficiency to be affected?

TX Don't go over 45 seconds
Temperature of Heat Shock, and time
Temperature bacteria is stored at
Incubation time and temperature

Part 2: Questions and Hypotheses

You have the opportunity to ask questions about bacterial transformation. Consider the protocols that you have researched. How might altering the steps affect transformation efficiency? Decide on a question, and then formulate a complete scientific hypothesis. Next lab, you can conduct an experiment that will enable you to compare the results of a standard protocol to a protocol that you've modified.

operate
How long does bacteria transformation take at longer incubation times

Write your question and hypothesis below:

IF bacteria growth improves with longer incubation time and we change the time from 10 mins to 20 mins then the bacteria should result in better growth.

Group 6

Part 3: Designing the Experiment

Regardless of your specific question, you will be working with three different treatment groups: a standard protocol, a modified protocol, and your negative control. Cells from your standard protocol and the protocol that you've modified will be inoculated on LB agar with ampicillin. Cells from the negative control, which will not receive plasmid DNA, will be inoculated on LB agar and LB agar with ampicillin. Predict whether there will be growth or no growth on each plate and determine the phenotype of cells from each treatment.

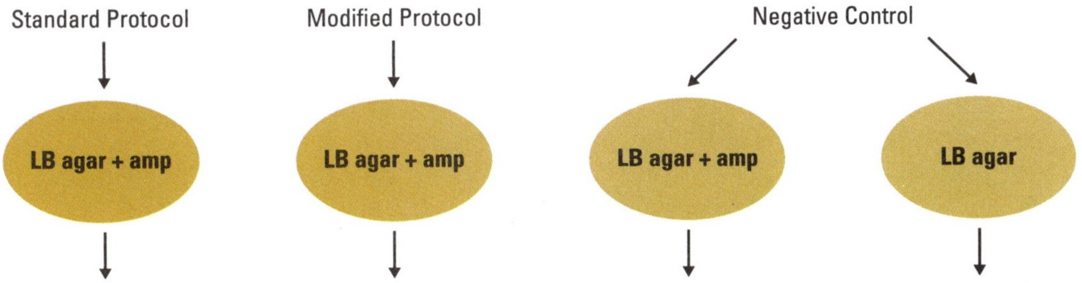

Growth: Normal Growth	Growth: Some Growth	Growth: No Growth	Growth: Normal Growth
Phenotype: Normal Growth	Phenotype: Some Glow	Phenotype: No Glow	Phenotype: NO Glow

Figure 1 Predicting traits. Determine whether or not cells will grow on plates and list the phenotype of cells from each treatment.

Consider the following as you design your experiment and prepare to carry it out.

Sterile Technique

It is critical that your work area, tools, and cultures stay clean as you work. Follow these tips for sterile technique.

- Only use sterile materials (autoclaved tips and tubes, flame sterilized cell spreaders, etc.).
- Do ***not*** reuse micropipette tips. Dispose of tips after each use.
- Keep lids and caps on all tubes and Petri dishes until you are ready to use them.
- Do ***not*** breath, sneeze, or cough into solutions and reagents.
- Wipe down workspace with disinfectant before and after use.

Pouring Agar Plates

You will grow your transformed bacteria on agar plates. Use sterile techniques to pour Luria broth (LB) agar plates. LB agar is a nutrient rich medium used to culture *E. coli*. The three plates that you will pour also contain the antibiotic ampicillin, which is used as a selectable marker. When pouring plates, it is important to use sterile techniques to prevent contamination. You don't want anything besides *E. coli* to grow on your plates. Follow the instructions below and those from your TA to pour agar plates.

1. Thoroughly clean your lab bench with disinfectant.
2. Obtain three sterile Petri dishes from your TA. **Prevent contamination—do not open the dishes until you are ready to pour.**
3. Label the bottom of your dishes with your section number and group name.

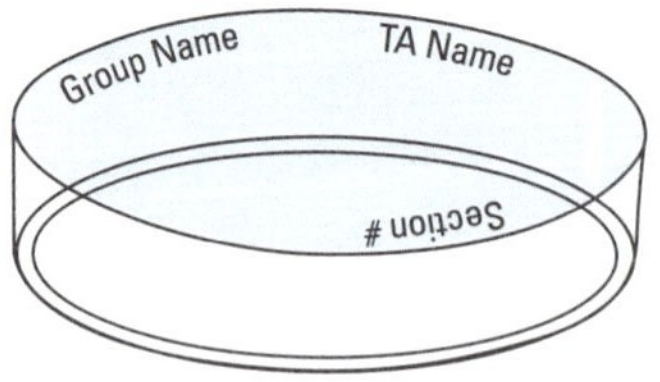

Figure 2

4. Using a Hot Hands Protector, obtain a flask of sterile LB + amp agar from the hot water bath. **Prevent contamination—do not open the lid until you are ready to pour.**

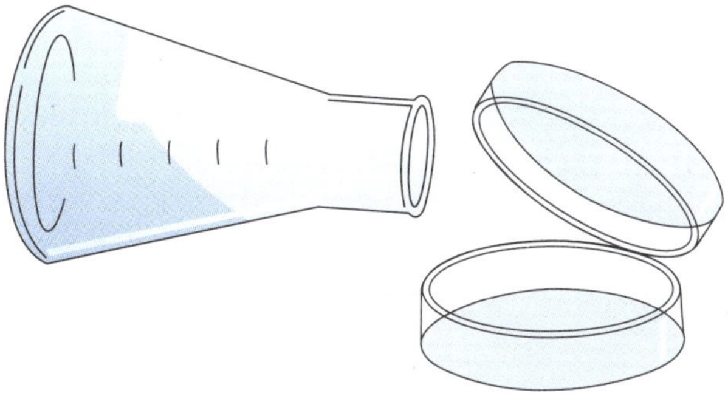

Figure 3

5. Lift one side of the Petri dish lid and pour enough media to cover the bottom of the plate. (Imagine that the lid and dish are hinged.)

6. Immediately close the lid on the dish.

7. Let your dishes sit on the lab bench undisturbed for 15 minutes. During this step, the media will cool, causing it to solidify into a gel.

8. Give your labeled plates to your TA to store until next week.

Transforming *E. Coli* Cells

Transforming *E. coli* cells is a three-step process. Follow the procedures to prepare chemically competent cells, perform the transformation, and culture the cells.

Step 1: Preparing Competent *E. Coli* Cells

Cells must be "competent" in order to be able to accept a plasmid. Follow the steps below to make "competent" *E. coli* cells.

1. Your TA will give you a microcentrifuge tube containing 1 mL of *E. coli* in LB broth.

2. Centrifuge the tube for two minutes at 14,000 RPMs to pellet cells.

3. Gently pour off the supernatant into a waste container. Take care not to disrupt the cell pellet.

4. Add 250 µL of 50 mM $CaCl_2$ to the cell pellet. **Mix solution via pipetting until the cell pellet has broken apart and solution is homogeneous.** Cap the tube.

5. Incubate the cell solution on ice for 10 minutes. During this step, the cells will become chemically competent. **For the remainder of the protocol, cells must be on ice!**

6. Label three microcentrifuge tubes as #1, #2, and #3. Place the closed, empty tubes on ice.

7. After the ten-minute incubation period, transfer 48 µL of cell solution (containing competent cells) into each of the three labeled tubes. Close the caps and keep tubes on ice.

Step 2: Transforming *E. Coli* Cells

Once you have chemically competent cells, follow the next set of steps to transform them. In this step, the plasmid will enter the bacterial cells. Fill in the table for the standard protocol, modified protocol, and the negative control.

	Tube #1 Standard Protocol	Tube #2 Modified Protocol	Tube #3 (–) Control
8. Add 2 μL of plasmid to 48 μL of cell solution for a final volume of 50 μL. Pipette the plasmid directly into the solution, not onto the side of the tube. Gently flick the tube to mix. ■ Specify the plasmid concentration for each tube. (Maximum plasmid concentration is 20 μg/μL) ■ Considerations: How could you modify the concentration given a 20 μg/μL stock? If you want to add 0 μg/μL plasmid, what will you add instead?	Solution: 2 μL 48 μL Concentration: 20 μL	Solution: 2 μL 48 μL Concentration: 20 μL	Solution: 1 μL Concentration: 0 μL
9. Incubate DNA/cell solution on ice. ■ Specify the duration of the incubation for each tube. (30 minutes maximum)	Time: 15 mins	Time: 20 mins	Time: 25 mins
10. Heat shock DNA/cell solution. ■ Specify the temperature and time for each tube. (5 minutes maximum)	Temperature: 42°C Time: 45 secs	Temperature: 42°C Time: 45 secs	Temperature: 42°C Time: 45 secs
11. Place DNA/cell solution on ice. ■ Specify the duration for each tube. (5 minutes maximum)	Time: 5 mins	Time: 5 mins	Time: 5 mins
12. Add 250 μL of media. ■ Specify the type of medium for each tube. (Available solutions: LB broth and SOC medium)	Medium: LB	Medium: LB	Medium: LB
13. Incubate cells for room temperature. ■ Specify the duration of the incubation for each tube. (10 minutes maximum)	Time: 10 mins	Time: 10 mins	Time: 10 mins

Step 3: Inoculating *E. Coli* onto Media

In this step, you will grow the bacterial cells in order to view the results of the transformation; you'll be able to determine if the bacterial cells received the plasmid and are expressing those genes. Once you have transformed your *E. coli* cells using the standard protocol, a modified protocol, and a negative control, you can plate your cells on agar to grow them. Once plated, they will incubate at 37°C for 24 hours. After the incubation time, lab staff will move them to a 4°C refrigerator until you view your plates next week.

14. Obtain **three** LB + amp plates and label them "amp1 standard," "amp2 modified," and "amp3 negative." Obtain one LB plate (no ampicillin) and label it "LB3 negative." Be sure your section number and group name are listed on all your plates.
15. Transfer 100 µL of cell solution from tube #1 onto the plate labeled "amp1 standard."
16. Sterilize the cell spreader by removing it from the ethanol beaker and allowing the tip of the spreader to touch the flame. After the flame burns out, let the spreader cool for 15 seconds. Be careful not to let anything touch the spreader and contaminate it. Don't blow on it or wave it around to cool it!
17. Gently move the spreader over the surface of the agar to distribute cells.
18. Place the spreader back in the ethanol beaker.
19. Repeat Steps **15–18** for the remaining samples, plating cells from the appropriate tubes onto plates.

Write out plans for your experiment below:

- Centrifuge a tube containing 1mL of E.coli at 14,000 RPM for 2 minutes.
- Gently pour off supernatant without disturbing the cell pellet.
- Add 250 µL of 50 mM $CaCl_2$ to the cell pellet and mix
- Incubate cell solution on ice for 10 minutes.
- Label 3 tubes #1, #2, #3 and place on ice.
- After incubation period, transfer 48 µL of cell solution into the 3 tubes.
- Look at figure in Step 2
-

Part 4: Carry out the Experiment

Once your TA has approved your plans, you can begin your experiment. Be sure to wear the proper personal protective equipment (lab attire, gloves, eye protection, etc.). As you conduct this experiment, please remember that you are working with *E. coli,* a human pathogen. Though this particular strain has been engineered to be less virulent, it is still an opportunistic pathogen. Always wear gloves when handling bacteria and dispose of any bacteria-contaminated items in the autoclave trash bins. As you work, remember that you are responsible for keeping the lab neat and tidy—a clean lab is a safe lab! Record data and any notes in the space below.

Use the space and table below to take notes and record data:

Part 5: Calculate Transformation Efficiency

The final part of this experiment is to analyze your data. Examine your bacterial plates. Count colonies that are expressing traits conferred by the pGLO plasmid. Calculate the transformation efficiency of your standard and modified protocols. The transformation efficiency is defined as the number of transformants per amount of DNA used. **Note:** When you inoculated bacteria onto your plates, you transferred 100 µL from a total volume of 300 µL. Therefore, you are only counting 33.33% of total number of transformants you obtained. Use the following equation to calculate transformation efficiency:

Transformation efficiency = (# transformants) / (_____ µg/µL of DNA × 2 µL × 0.33)

Write your transformation efficiency calculations in the space below:

Clean Up

1. Give your Petri dishes to your TA to store until next week.
2. Dispose of all bacterial-contaminated items in the autoclave trash bin.
3. Clean your lab bench with disinfectant.
4. Wash your hands!

Assignments

- In-Class
  - Take good notes in your lab manual. Actively participate in lab.
- Homework
  - Answer the Transformation Questions. Submit the assignment using Turnitin on eCampus. This is an independent homework assignment.
  - Complete a Testing Hypotheses Worksheet using notes you took in lab. This worksheet is based on the experiment you design and carry out. Type it using the electronic form for the assignment. Submit it through the Turnitin link on eCampus. This is an independent homework assignment.
  - Prepare a presentation according to your TA's instructions.
  - Read Lab 7: Science of Opioid Dependence
  - Quizzes

Transformation Questions

Answer the following questions using complete sentences and well thought out paragraphs that completely and clearly answer the questions. For this assignment you may consult primary or secondary sources. While you are not required to submit a formal reference section, do include links to websites you've used to help you answer the questions. Each person is required to write answers to the questions on his/her own. Submit the assignment using Turnitin on eCampus.

1. Describe the process of a bacterial transformation. Explain how scientists can make bacteria take up DNA in a laboratory setting.

2. Genetic transformations are not limited to bacteria. There are real-world applications where scientists have taken genes from one organism and inserted them into another organism. Give an example of how a genetically modified organisms (can be any organism, you are not limited to bacteria) have been used to solve real-world issues. Describe the organisms and gene(s) that were manipulated. Explain how the transgenic organism is beneficial to us (i.e., Does it help us produce better medicine? Does it help crops grow faster? Does it help decrease pollution? etc.).

3. Come up with a novel question/problem that you think could be answered/resolved with the application of a genetic transformation. Include details: What organism would you genetically modify and why? How would your genetic modification alter the phenotype or physical characteristics of the organism? How would this transformant answer/resolve your question/problem?

Presentations

To demonstrate what you have learned through this lab, you will be giving a presentation. A portion of the presentation will be based on your experiment from the Bacterial Transformation Lab. Other portions may be based on library or Internet research. Your TA will provide more information about expectations and grading for the presentation in lab.

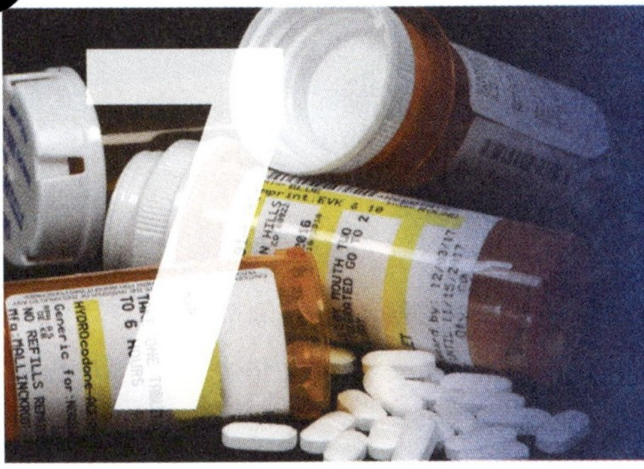

Lab 7 | Science of Opioid Dependence

Assignments

In-Class

- ❒ Notes and participation

Homework

- ❒ Complete a Testing Hypotheses Worksheet following the instructions discussed in lab. Submit the assignment using Turnitin on eCampus.
- ❒ Complete and submit the Science of Opioid Dependence homework assignment.
- ❒ Quizzes

Learning Goals

By the end of this lab, you will be able to:

- Understand how opioids affect neurotransmitters in the brain
- Have learned about tools scientists use to study opioid dependence
- Use gel electrophoresis to study DNA samples
- Make conclusions about possible links between genetics and opioid dependence

INTRODUCTION

Opioids are in news headlines daily. There has been a steady number of people becoming dependent on opioids since the mid 1990s, and today there are record numbers of people dying due to opioid overdoses (HHS 2019). West Virginia is among states that have been hardest hit by the opioid epidemic and has the highest opioid-related death rate in the country (Washington Post 2019). In 2017, the Health and Human Services secretary declared a public health emergency to address the national opioid crisis and included a 5-point strategy to combat the crisis. One of the priorities of that strategy is "strengthening our understanding of the epidemic" (HHS 2019). With this lab, you will learn about some approaches being taken by researchers and you will work through a study aimed to help address the opioid epidemic.

Natural opioids, like endorphins, activate reward pathways in the brain. They bind to receptors that cause the release of dopamine, a neurotransmitter, which results in feelings of well-being and possible pain reduction. Synthetic opioids, like oxycodone or heroin, mimic endorphins. Synthetic opioids may be prescribed to patients to treat severe or long-term pain, such as pain from surgery or chronic illness.

With high doses or prolonged use, synthetic opioids can become dangerous and have serious side effects, including respiratory arrest possibly leading to death. Patients taking opioids are also at risk of developing dependence; their bodies build up a tolerance to the drug and need higher doses to achieve the same effect. A person who is dependent on opioids will experience withdrawal symptoms when not exposed. It is important to point out that a person who is dependent on opioids will not necessarily abuse drugs or develop an addiction. Someone who has an addiction exhibits behaviors that include compulsive urges to seek out and use the drug.

Given the risks of opioid dependence and abuse, doctors must use caution when prescribing opioids for pain. There are screening tools they can use that are based on collecting a person's health history and other environmental factors that have been indicated in drug abuse risk. These screening tools have their limitations, so some researchers are looking for other factors related to opioid dependence, like genetics.

Two genes on chromosome 11 are involved in opioid reward pathways and are targets of studies investigating possible genetic links to opioid dependence. These two genes are the DRD2 (D2 dopamine receptor) gene and the ANKK1 (ankyrin repeat and kinase domain containing 1) gene. Within these genes there are single nucleotide polymorphisms or SNPs (pronounced "snips"). These single base pair differences represent different alleles of the gene in the population. In this lab, we will study the SNP rs1800497, which has a C allele and a T allele.

Scientists can determine someone's genotype for rs1800497 by taking advantage of a restriction site found in the C allele, but not the T allele. In the assay, the rs1800497 region of an individual's genome is isolated using DNA primers and amplified, or copied many times, using the polymerase chain reaction (PCR). Then the DNA PCR product is digested with the restriction enzyme TaqI. TaqI will cut the PCR products of the C allele, but will not cut PCR products of the T allele. Gel electrophoresis can then be used to analyze the sizes of the DNA fragments. The bands that show on the gel are indicative of a person's genotype (CC, TT, or CT).

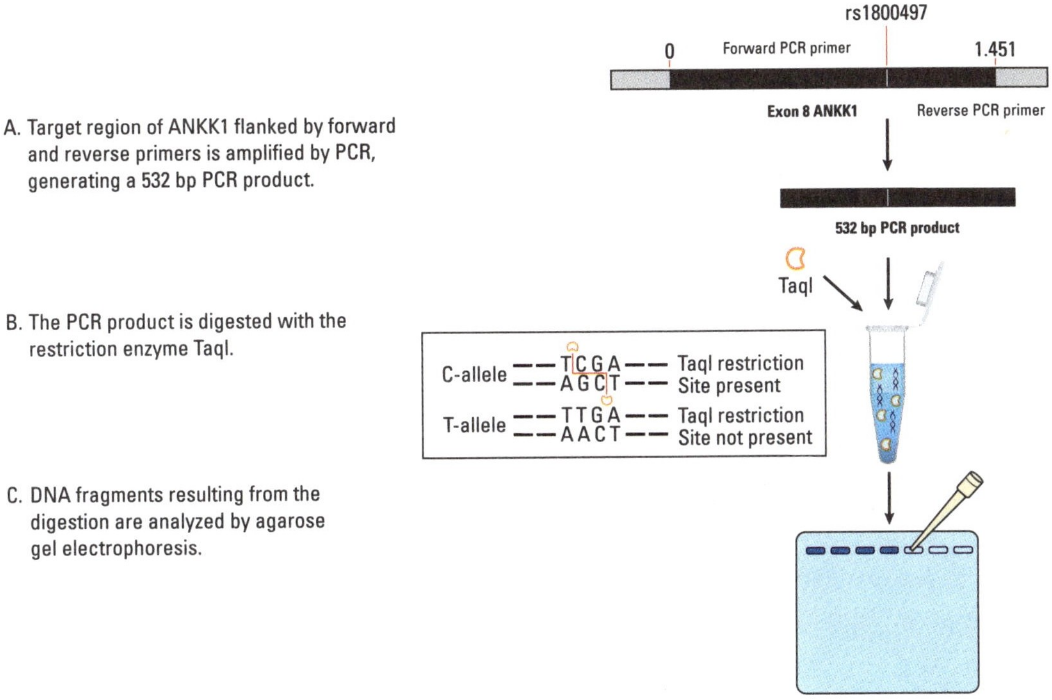

Figure 1

References

Cenziper D, Corio D, Hooper K, Soule D. 2019 Oct 18. The Opioid Files: They looked at us like an easy target. Washington Post. [accessed 2019 Nov 1]. Available from: https://www.washingtonpost.com/graphics/2019/investigations/west-virginia-opioid-legal-battle-foster-care/

Department of Health and Human Services: National Opioids Crisis. 2019. Washington, DC; [cited 2019 Nov 1]. Available from: https://www.hhs.gov/opioids

Part 1: Studying Opioid Dependence

1. How do opioids affect the brain?

They release dopamine and also reduces pain.

2. What are some screening tools researchers and medical professionals use to identify risks of opioid dependence or abuse in patients?

Rehab patients
Genetics

3. What are limitations of surveys and questionnaires? Why do you think researchers are interested in genetic elements of susceptibility to opioid dependence?

Some of them can talk about risks at differant ages & enviroment. Can't expect honesty.

Part 2: Planning a Study

You are part of a research team investigating genetic links to opioid dependence. You have the ability to sample a control population that represents individuals from the general population who are not dependent on opioids, and a case population who are individuals who have been identified as opioid dependent.

1. What is your research question?

What is the allele that shows more frequent in an individual identified as opoid dependent than individuals who are not?

2. Describe the attributes of your case and control groups. What will be different between them? What will they have in common?

2 groups of individuals. One opoid dependent.
One opoid independent.
Alleles

3. What will you measure and compare?

Measuring Frequency of alleles

4. Outline a plan for your research study. Include (brief) methods you'll use to study genes and alleles. You'll have access to:

   a. DNA samples that are PCR products amplified from the SNP rs1800497 locus and treated with TaqI.
   b. Molecular weight marker with various DNA sizes.
   c. Control DNA from an individual with CT genotype.
   d. Agarose gels, running buffer, dyes, and light for visualization.
   e. DNA electrophoresis units.

(H) If allele frequency is different between genes DRD2 & ANKK1 and you compare the C allele frequency in both opoid dependent and non-dependent individuals than you shoud see more of the ANKK1 gene in opoid dependent individuals.

5. Make a sketch of a gel showing individuals with CC, TT, and CT genotypes.

Part 3: Carrying out the Study

Gel Electrophoresis

Gel casting apparatus parts

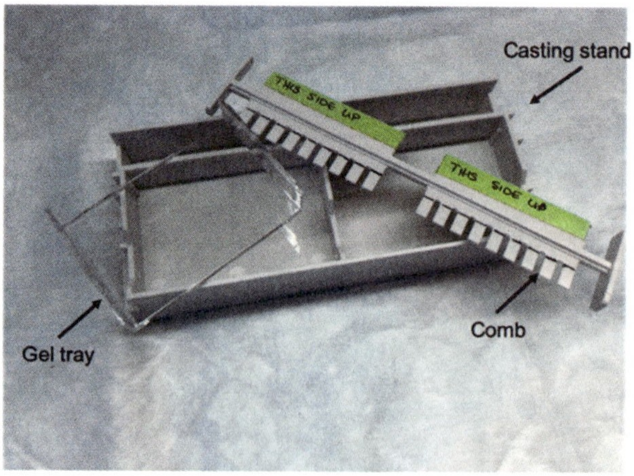

Figure 2 Gel casting apparatus. You should have the parts shown.

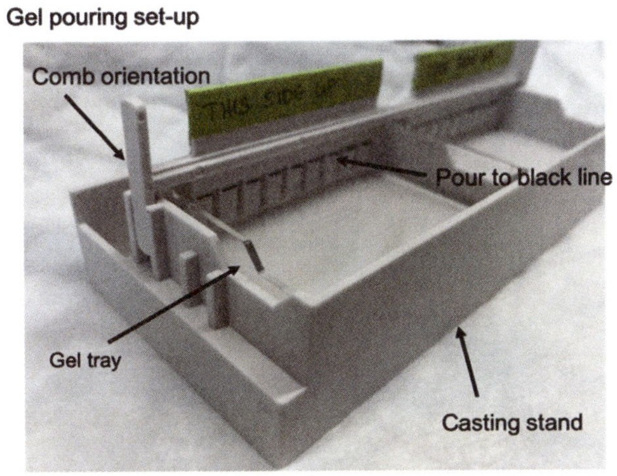

Figure 3 Gel pouring setup. Please note the location and orientation of the comb.

Pouring an Agarose Gel

1. Refer to Figure 2 for a picture of the gel casting apparatus and ensure you have the required parts to pour the gel.

2. Select a member of the group to pour the gel. (That individual should wear gloves and eye protection when pouring.)

3. Place the gel tray into the casting apparatus. The tray fits in only one orientation. It should slide into place easily. ***Do not*** force the gel tray into the casting box.

4. Insert the gel comb into the slots (see Figure 3). ***Do not*** force the comb into the slots. The comb will form wells where you will load your samples.

5. Make sure the gel apparatus is on a level surface and that the tray and comb are properly in place.

6. Carefully pour 20 mL of a 1% TAE agarose solution containing SYBR Safe DNA gel stain into the gel tray. The gel should be approximately 1 mm from the top of the comb's teeth. (Look for the line on the comb indicating the correct level of agarose.) Your TA can assist you with seeing if you have poured enough, but not too much, agarose.

   - **Note:** Some leakage of agarose under the gel tray is normal.

7. Do not move the gel apparatus until the gel has completely solidified. This takes about 10–20 minutes. It will appear uniformly cloudy.

Loading the Gel

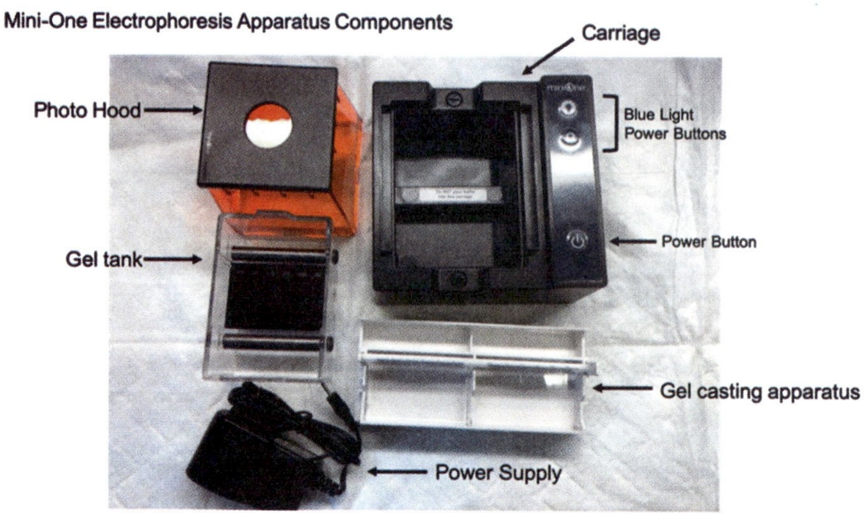

Figure 4 Mini One Electrophoresis Apparatus. Please note the names of the parts.

1. Refer to Figure 4 for a picture of the electrophoresis apparatus and ensure you have the required parts to set up and run the gel.

2. Insert the ***dry*** gel tank into the carriage by aligning the negative and positive terminals. Make sure it snaps into place, but do not force the gel tank into place. If any moisture remains on the components from prior use, carefully dry them with a paper towel or kimwipe. ***Do not*** allow the carriage to get wet as this may permanently damage the electrophoresis apparatus!

3. When the gel has solidified, gently pull the comb straight up.

4. Remove the gel tray with the solidified gel from the casting apparatus and wipe any excess gel from the outside of the gel tray.

5. Gently place the gel tray with the gel into the gel tank so the wells are oriented closest to the negative terminal. (When the gel is running, DNA samples loaded into the wells will migrate toward the positive terminal.)

6. Pour 1× TAE buffer very slowly into one end of the gel tank until the buffer covers the gel (approximately 140 mL). The top of the gel should be covered by about 2 mm of buffer. ***Do not*** overfill the chamber or allow buffer to leak outside of the gel tank as this will permanently damage the electrophoresis apparatus!

7. Attach the power supply and plug in the apparatus. Check to make sure that the electrophoresis apparatus will turn on by securely seating the photo hood on the top of the carriage then pressing the power button on the apparatus. A solid green LED light should be visible. This confirms that there is enough buffer and the apparatus will run. If the green light does not turn on, ask your TA for assistance. You may need to adjust the amount of buffer in the gel tank or reseat the gel tank within the base.
   - **Note:** The apparatus will not turn on until the photo hood is seated correctly.
8. Turn off the unit (no green light) so you can safely load samples.
9. Turn on the blue light (light bulb symbol). There are two intensities of light. Use whichever one helps you see the wells the best.
10. Using a fresh tip for each sample, load the molecular weight marker and samples according to the table below.

Table 1 Loading Volumes for Electrophoresis Controls and Samples

Lane	Sample	Volume
1	Molecular Weight Marker	10 µL
2	CT PCR control	10 µL
3	Sample 1 (Tube #A2165)	10 µL
4	Sample 2 (Tube #A3022)	10 µL
5	Sample 3 (Tube #A2047)	10 µL
6	Sample 4 (Tube #B1027)	10 µL
7*	Optional: Sample 5 (Tube #__________)	10 µL
8*	Optional: Sample 6 (Tube #__________)	10 µL
9		
Be sure to fill in your tube ID Number! It is necessary for future calculations.		

***Please note:** Number of student samples may change depending on how many members are in your group. Each group member should load a sample.

Wells on negative side

Running the DNA Samples

1. Securely seat photo hood onto the carriage.

2. Turn on the electrophoresis apparatus by pushing the power button. A solid green light should be lit.

3. Set a timer and allow the gel to run for 20 minutes. The electrophoresis apparatus is designed to run at a constant voltage of 42V when on (green light is lit). If the green light is not on, the unit is not turned on or there is a problem with your apparatus. It must remain lit the entire duration of the run.

4. Check the progress of the run occasionally throughout the 20 minutes. You should see yellowish–green bands moving toward the positive electrode and separating out within the gel.

5. After 20 minutes, turn off the power to the unit (green light is off).

6. One group member should remove their gloves and wash their hands so that they may take a picture of the gel using a cellphone and share it with the group. Rest the camera lens of the phone against the top of the box—this is the perfect height for the picture to be in focus and capture all parts of the gel. You should also take notes about the gel appearance and number of bands in each lane, etc. in your lab notes. Be sure you have a picture before you clean up! The gels will not be saved.

Cleaning Up

1. After you have taken notes about the gel and taken a picture for your group, carefully remove the gel tray from the gel tank and dispose of the gel into the regular trash. ***Do not throw away the transparent gel tray.*** Rinse the gel tray with cool water and allow to dry.

2. Carefully remove the gel tank from the carriage and dispose of the electrophoresis buffer down the sink drain. Rinse the gel tank with cool water and turn upside down onto paper towels to dry. If any liquid remains on the carriage or photo hood, gently use a paper towel to dry them.

3. Throw away any micropipette tips and other waste used during the experiment in the regular trash.

Part 4: Data Analysis

Practice with Data

Using numbers provided by your TA, practice calculations to compare control and case populations.

Table 2 Genotypes of Individuals

Genotype	Number in Control Population	Number in Case Population
CC	245	180
CT	220	320
TT	65	115
Total:	530	615

Once you have the numbers of individuals with each genotype, use the formula below to calculate the genotype frequencies and record the data in Table 3.

$$\text{Genotype frequency} = \frac{\text{\# individuals with genotype (CC, CT, or TT)}}{\text{total \# individuals}}$$

Table 3 Genotype Frequencies

Genotype	Frequency in Control Population	Frequency in Case Population
CC	.46	.29
CT	.42	.52
TT	.12	.19
Total:	1	1

(2 × 65) + 220 (2 × 180) + 320

(2 × 245) + 220 (2 × 115) + 320

Using the data in Table 2, calculate the number of alleles and the allele frequencies using the formula below and record the data in Tables 4 and 5.

Alleles = (2 × # individuals with homozygous genotype (CC or TT)) + (# individuals with heterozygous genotype (CT))

Table 4 Numbers of Alleles

Alelles	# in Control Population	# in Case Population
C	710	680
T	350	550

$$\text{Allele frequency} = \frac{(2 \times \#\text{ individuals with homozygous genotype (CC or TT)}) + (\#\text{ individuals with heterozygous genotype (CT)})}{2 \times \text{total }\#\text{ individuals}}$$

Table 5 Allele Frequencies

Alelles	Frequency in Control Population	Frequency in Case Population
C	.67	.55
T	.33	.45

Statistical Analyses

Statistical analyses, such as a chi square test, could help you determine if any differences you observe between control and case populations are real (significant) or occurring by chance. One output of chi square analysis is a probability value, or *p*-value. These *p*-values are a measure of how likely it is the differences happened by chance.

A *p*-value of less than 0.05 is usually considered significant and means that 19 times out of 20 the difference or relationship between the identified data points are real and only one time out of 20 would the difference occur by chance. Also, while a *p*-value of 0.05 or less is considered significant, *p*-values can go much lower, indicating higher significance the lower they go. So a *p*-value 0.0001 is much more significant than 0.05 and offers stronger evidence that the difference is real.

There are online calculators available for performing chi square tests. Go to *https://www.socscistatistics.com/tests/* and select the chi square calculator for 5 × 5 (or less) contingency table to perform analyses for genotype and allele frequencies by entering categories (genotypes or alleles) and groups (control and case populations), and numbers of individuals in the calculator. Use the results to determine if there are significant differences between control and case populations.

Table 6 *p*-Values From Chi Square Analysis

	p-Value from Chi Square Analysis	Significant Difference?
C	<.00001	<.00001
T	<.00001	<.00001

Analyzing the Collected Data

Reading the Gel

In Table 7, record the results of the gel by indicating whether bands that would indicate the C and/or T allele are present in each lane.

Table 7 Presence of C/T bands in Samples

Lane	Sample	C band?	T band?
1	Molecular Weight Marker	———	———
2	CT PCR control	———	———
3	Sample 1 (Tube # A2105)	———	———
4	Sample 2 (Tube # A3022)		———
5	Sample 3 (Tube # A2047)	———	———
6	Sample 4 (Tube # B1027)	———	
7*	Optional: Sample 5 (Tube #________)		
8*	Optional: Sample 6 (Tube #________)		
9			

Calculating Genotype and Allele Frequencies

Fill in Table 8 with the sample information. The samples were de-identified to protect patient privacy. Your TA will have a key to help you determine whether you had a sample from the control population (non-opioid dependent) or case population (opioid-dependent). Once you have identified your samples, your TA will have you add your data to the class data.

Table 8 Sample Identification

Sample	ID #	Control or Case?	Genotype
1		N/A	CT
2		N/A	CT
3	A2105	Control	CT
4	A3022	control	TT
5	A2047	control	CT
6	B1027	case	CC

Once the class data has been collected, complete Table 9 with the information.

Table 9 Genotypes of Individuals for Your Lab Section

Genotype	Number in Control Population	Number in Case Population
CC	2	2
CT	8	7
TT	2	3
Total:	12	12

Once you have the numbers of individuals with each genotype, use the formula below to calculate the genotype frequencies and record the data in Table 10.

Table 10 Genotype Frequency of Patients for Your Lab Section

Genotype	Frequency in Control Population	Frequency in Case Population
CC	.17	.17
CT	.66	.58
TT	.17	.25

Using the data in Table 9, calculate the number of alleles and the allele frequencies using the formula below and record the data in Tables 11 and 12.

Alleles = (2 × # individuals with homozygous genotype (CC or TT)) + (# individuals with heterozygous genotype (CT))

Table 11 Numbers of Alleles for Individuals for Your Lab Section

Alleles	# in Control Population	# in Case Population
C	12	11
T	12	13

$$\text{Allele frequency} = \frac{(2 \times \text{\# individuals with homozygous genotype (CC or TT)}) + (\text{\# individuals with heterozygous genotype (CT)})}{2 \times \text{total \# individuals}}$$

Table 12 Allele Frequency for Individuals for Your Lab Section

Allele	Frequency in Control Population	Frequency in Case Population
C	.5	.45
T	.5	.55

Using online chi-square calculator, compare genotype and allele frequencies for your lab section. Indicate the *p*-value and determine if there are significant differences between your control and case samples in your lab section.

Table 13 *p*-values From Chi Square Analyses

	p-Value from Chi Square Analysis	Significant Difference?
C	.772638	.772638
T	.772638	.772638

Assignments .875173

- In-Class
  - Take good notes in your lab manual. Actively participate in lab.
- Homework
  - Researchers are conducting studies to learn more about possible contributions of genetics to opioid dependence. Find a primary research article related to genes, opioids, and dependence. Use the article to complete a Testing Hypotheses Worksheet. This is an independent homework assignment.
  - Complete and submit the Science of Opioid Dependence Homework Assignment. Submit the assignment using Turnitin on eCampus. This is an independent homework assignment.
- Quizzes

Science of Opioid Dependence Assignment

A template for this assignment can be found on eCampus. Type your answers to the following questions into the document and submit it through Turnitin links on eCampus by the due date.

1. What was the question you wrote for this research study? (3 pts.)
2. What particular region of DNA was isolated and amplified for this study? Why was it selected? What was unique about this region? (4 pts.)
3. Briefly describe the steps or methods for studying genes. Start with targeting a gene region and end with gel electrophoresis. (4 pts.)

4. How did using gel electrophoresis help you determine the genotypes of individuals? Incorporate an explanation of how gel electrophoresis works in your answer. Include a picture of your gel to help you explain. (4 pts.)

5. Your TA will provide you with numbers collected across lab sections in the course. Compare those numbers with the ones collected by your lab section.

   a. Calculate genotype and allele frequencies for your lab section and across lab sections in the course. (3 pts.)

   b. Explain how your lab section compares to combined lab sections. Include tables to summarize the data with your explanation. (3 pts.)

   c. Do chi square analyses to determine if there are significant differences between control and case populations with regard to genotype and allele frequencies for your lab section and combined lab sections. (3 pts.)

   d. Explain how your lab section compares to combined lab sections in the course. Provide summary results of chi square tests with your explanation. (3 pts.)

   e. What could account for differences between your lab section and across lab sections in the course? Do you think one data set is more reliable than the other? Explain. (3 pts.)

6. What conclusions can you make about the case and control populations. Is there a genetic association to opioid dependence? (4 pts.)

7. Why is it inaccurate to say that if someone has a T or C allele for the rs1800497 SNP they will become dependent on opioids? (4 pts.)

8. Current published research on the ANKK1 gene and the SNP rs1800497 showed that the frequency of the T allele was 38% in a control group and 44% in a group of opioid dependent participants in a study (Zhang et al. 2018). Chi square analysis resulted in a p-value of 0.04. Do your results align with published data? Explain. (4 pts.)

9. How do you think the methods for this lab (i.e. studying genes and DNA) could be used to address the opioid epidemic? (4 pts.)

10. Thinking more broadly than in the last question, what do you think could be done to address the opioid crisis? List at least two ideas you have. Explain why you think your suggestions would be effective. (4 pts.)

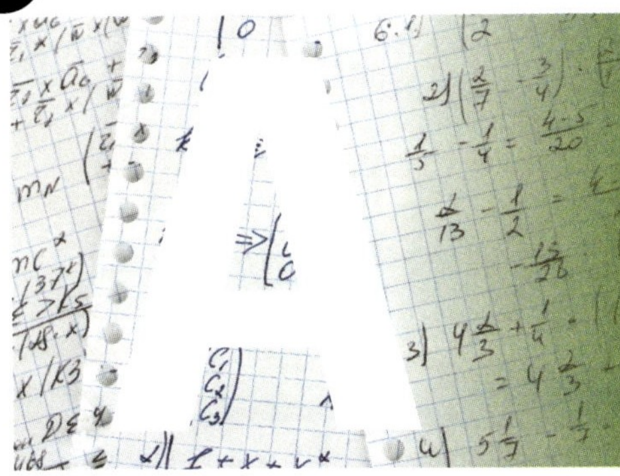

Appendix A | Communicating in Science—Writing Lab Reports in the Style of a Scientific Journal Article

Scientific Communication—Using PACT Principles

One of the most important skills you can learn as a young scientist is how to communicate well. Scientists in all disciplines communicate their findings to professional audiences. They often do this by publishing scientific papers in peer-reviewed journals. Throughout the semester, you'll be writing sections of lab reports in the style of scientific journals. These assignments will help you hone your writing and communication skills.

The Eberly College SpeakWrite program uses PACT principles to enhance communication and writing skills. Consider these four principles when writing lab reports in the style of scientific journals.

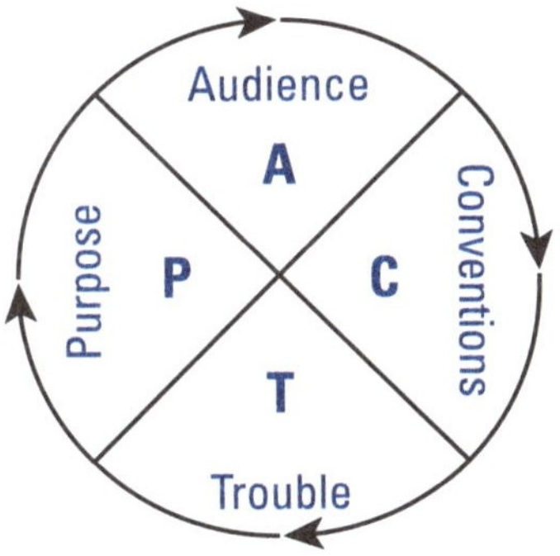

Purpose: What exactly do I want to happen?
Audience: Who is reading, listening, or viewing?
Conventions: What is expected in this context?
Trouble: What could get in the way of my goals?

The main **purpose** of writing scientific journal articles (and lab reports) is to describe your research, put it in context of what it already known, and discuss future implications of the work. For you as a biology student, you can use your lab report to demonstrate your understanding of your research. For research scientists, publishing papers demonstrates that they are doing high quality experiments and are worthy of grant funding and promotion.

Scientific journal style papers are written for an **audience of peers.** For you as a biology student, that means writing to someone who has a basic understanding of biology. Research scientists will write papers for professional audiences. Scientific papers are typically written using scientific terminology (jargon) because it is assumed that the reader already has a background in the subject.

The **conventions** of scientific journal style papers can vary depending on the particular journal, but typically they include similar sections: introduction, methods, results with figures, discussion, and references. They are written formally, without flowery language or casual tone. Additionally, the writing should be clear, concise, and objective. Finally, experiments in journal style papers are put into context by referring to previous work using citations. While it is not appropriate to quote previous work, it is expected that the author will paraphrase previous findings that give their own work meaning.

The **trouble spots** in scientific journal style writing are numerous. However, the first things that new scientists can do to avoid trouble spots is to understand the expectations of this type of writing. That means, having goals in mind for the paper and each section of the report, directing the writing to the appropriate audience, and making sure that sections of the lab report contain the right kind of information. All of this takes time and practice; be sure to leave plenty of time for proofreading, editing, and asking questions when you aren't sure!

Sections of a Scientific Paper

By convention, scientific papers are separated into sections. Below is a brief description of each section.

Title Section

The title should be concise, but descriptive. Your reader should be able to infer your treatments and what you measured (your independent and dependent variables). A good title often even hints at the results of the experiment. A scientific paper will list collaborators and affiliation—for you this means including your lab partners' names as well as section information and your TA's name.

Introduction

The introduction of a scientific paper introduces the reader to the topic by explaining the rationale for the experiment. In other words, it explains why the question is important or relevant. Background information is included to help familiarize the reader with the topic. Citations are used in this section to indicate what is already known about the topic and show how the new questions fit with the current knowledge. This sections starts with general background information, then narrows to focus on the specific question being asked, and ends with the hypothesis.

Methods

The methods section should include a brief, but complete narration of methods that you used in your experiment. It is written in past tense and should not be written as a list of instructions. Level of detail is important—you should include enough information so that someone could repeat your experiment, but avoid unnecessary details. The reader should be able to infer components of your experimental design (treatments, variables, controls, constants, replication, etc.). The methods you use must be appropriate to test the hypothesis you posed at the end of the introduction.

Results

The results section is a written summary of trends of your data and should highlight the important findings. Include the results of statistical tests and comparisons. This section is written using complete sentences. There is no interpretation of the data in this section, meaning there is no need to explain why you think you got the results or what the results might imply—save this for the Discussion. Figures and tables are included with the results to visually summarize the data for the reader. Figures and tables should be accurately labeled and have informative legends or captions. References to figures and tables are included in the text of the results.

Discussion

The discussion section is the place to indicate whether or not the hypothesis was supported, then explain how your findings are relevant to the bigger picture. It is in this section where you interpret the data for the reader and explain the implications of the research findings. This section will have citations that allow the author to compare findings with others' work published in the scientific literature. It is also the place to give alternative explanations for findings, state any limitations of the experimental design, and indicate future direction of research.

References

This section is a list of all references used for the lab report. If you cite a source in the text, it must be listed here, and if you list a reference here, it must be cited in the text. References are formatted using the CSE name-year style.

Lab Report Rubric

Criteria for a good lab report written in the style of a scientific journal are outlined in the rubric on the next few pages. Use relevant portions of the rubric based on which sections have been assigned. Refer to the rubric as you write—you will receive feedback and a score based on your level of performance.

Criteria	Not Addressed	Novice	Intermediate	Exemplary
Title Section	(0)	(0.5)	(1)	(2)
Title is a clear, concise, informative description. Name of author Name of instructor/TA Course and section information	▪ Missing title or name and any of the following: day and time class meets, TA's name	▪ Title and name present. ▪ Missing any of the following: day and time class meets, TA's name ▪ Title is insufficient by not being descriptive of the report.	▪ Title and name are present. ▪ Missing one of the following: day and time class meets, TA's name ▪ Title isn't clear or concise.	▪ Title and name are present. ▪ All other parts present day and time class meets, TA's name ▪ Title is a clear and concise statement that describes what lies in the report.
Introduction: Context	(0)	(1–4)	(5–6)	(7–8)
Demonstrates a clear understanding of the big picture; Why is this question important/ interesting in the field of biology?	▪ The importance of the question is not addressed. ▪ How the question relates within the broader context of biology is not addressed.	▪ The writer provides a generic or vague rationale for the importance of the question. ▪ The writer provides vague or generic references to the broader context of biology.	▪ The writer provides one explanation of why others would find the topic interesting. ▪ The writer provides some relevant context for the research question(s).	▪ The writer provides a clear sense of why this knowledge may be of interest to a broad audience. ▪ The writer provides ample relevant context for the research question(s).
Introduction: Accuracy and Relevance	(0)	(1–3)	(4–5)	(6–7)
Content knowledge is accurate, relevant and provides appropriate background for reader, including defining critical terms.	▪ Background information is missing or contains major inaccuracies. ▪ Background information is accurate, but irrelevant or too disjointed to make relevance clear. ▪ References are absent, irrelevant, or are websites.	▪ Background omits information or contains inaccuracies which detract from the major point of the paper. ▪ Background information is overly narrow or overly general (only partially relevant). ▪ References are inadequately explained.	▪ Background information may contain minor omissions or inaccuracies that do not detract from the major point of the paper. ▪ Background information has the appropriate level of specificity to provide relevant context. ▪ References are relevant and adequately explained, but few.	▪ Background information is completely accurate. ▪ Background information has the appropriate level of specificity to provide concise and useful context to aid the reader's understanding. ▪ References are relevant, adequately explained, and indicate a reasonable literature search.
Introduction: Hypotheses	(0)	(1–2)	(3–4)	(5)
Hypotheses are clearly stated, testable; Hypotheses have scientific merit.	▪ No hypothesis is indicated. ▪ Hypothesis is trivial, obvious, not plausible, not testable, or is completely off topic.	▪ The hypothesis is present but poorly stated or too vague or confused for its value to be determined. ▪ Hypothesis is stated but not testable.	▪ The hypothesis is present and testable, but not clearly stated, or taken directly from course material.	▪ A relevant, testable hypothesis is clearly stated and indicates a level of understanding beyond the material directly provided to the student

Criteria	Not Addressed	Novice	Intermediate	Exemplary
Methods: Overall	(0–1)	(2–5)	(6–8)	(9–10)
Describes how the study was conducted with appropriate level of detail; Someone could repeat the experiment; Written in the past tense.	■ List of steps, not in sentence form. ■ Out of sequence, missing, or confusing.	■ List of instructions, or are not written in sentence form. ■ Some steps are clear; most are confusing or lack sufficient detail.	■ Written in sentence form, past tense. ■ Most steps are clear with only a few that are confusing or lacking sufficient detail.	■ Written in sentence form, past tense. ■ Presents easy-to-follow methods, which are logical and adequately detailed.
Methods: Experimental Design	(0–1)	(2–5)	(6–8)	(9–10)
The experimental design is likely to produce salient results; The experiment is designed to test the hypotheses posed.	■ Experimental design is inappropriate to test hypotheses. ■ Design is poorly explained or is indecipherable.	■ Experimental design is appropriate to test hypotheses. ■ Design details (independent and dependent variables, constants, controls, and replication) are vaguely explained.	■ Experimental design is appropriate to test hypotheses. ■ Design details (independent and dependent variables, constants, controls, and replication) are explained, but with minor omissions or mistakes.	■ Experimental design is appropriate to test hypotheses. ■ Design details (independent and dependent variables, constants, controls, and replication) are clearly explained.
Results	(0)	(1–4)	(5–6)	(7–8)
Presentation of data should be written in complete sentences. Data are comprehensive, accurate, and relevant.	■ Data and trends are not presented. ■ Results are not written in sentence form or are too haphazard or incomplete to be useful to the reader. ■ Results of statistical tests or comparisons are not presented.	■ Some necessary data and trends are missing or inaccurate. ■ Results are not written in complete sentences, lack organization, or lack figure references. ■ Results of statistical tests (i.e., *p*-values) or comparisons are presented, but contain major inaccuracies, errors, or are not consistent with data presented in figures.	■ Data and trends presented are relevant, accurate, and complete, with any gaps being minor. ■ Narrative of results is written in complete sentences, with only minor issues with organization or references to figures. ■ Results of statistical tests (i.e., *p*-values) or comparisons are presented with trends, but contain minor inaccuracies or inconsistencies with data presented in figures.	■ Data and trends presented are relevant, accurate and comprehensive. ■ Narrative of results is written in complete sentences, is logically organized, and has references to figures. ■ Results of statistical tests (i.e., *p*-values) or comparisons are accurately presented with trends, and are consistent with data presented in figures.

Criteria	Not Addressed	Novice	Intermediate	Exemplary
Results: Figures and Tables	(0–1)	(2–6)	(7–10)	(11–12)
Data are summarized in a logical format. Table or graph types are appropriate. Data are properly labeled including units. Graph axes are appropriately labeled and scaled and captions are informative and complete.	■ Relevant data is not presented in appropriate figures and/or tables. ■ Labels or units are missing which prevent the reader from being able to derive any useful information from the graph. ■ Presentation of data is in an inappropriate formats or graph types. ■ Captions or figure legends are confusing or indecipherable.	■ Some data is presented in appropriate figures and/or tables, but portions of relevant data are missing. ■ Contains some major errors in or omissions of labels, scales, units etc., but the reader is able to derive some relevant meaning from each figure. ■ Is technically correct but inappropriate format prevents the reader from deriving meaning or using it. ■ Captions or figure legends are missing or inadequate.	■ Relevant data is presented in appropriate figures and/or tables with only minor omissions. ■ Contains only minor mistakes in labels, scales, units etc. that do not interfere with the reader's understanding and the figure's meaning is clear without the reader referring to the text. ■ Graph types or table formats are appropriate for data type. ■ Includes captions or figure legends that are at least somewhat useful.	■ All relevant data is presented in appropriate figures and/or tables. ■ Contains no mistakes. ■ Uses formats or graph types that highlight relationships between the data points or other relevant aspects of the data. ■ Has informative, concise and complete captions or figure legends that include a title.
Discussion: Conclusions	(0–1)	(2–5)	(6–8)	(9–10)
Conclusion is clearly and logically drawn from data provided. A logical chain of reasoning from hypothesis to data to conclusions is clearly and persuasively explained. Conflicting data, if present, are adequately addressed.	■ Conclusions have little or no basis in data provided. ■ Connections between hypothesis, data and conclusion are nonexistent, limited, vague or otherwise insufficient to allow reasonable evaluation of their merit. ■ Conflicting data are not addressed.	■ Conclusions have some direct basis in the data, but may contain some gaps in logic or data or are overly broad. ■ Connections between hypothesis, data and conclusions are present but weak. ■ Conflicting or missing data are poorly dressed.	■ Conclusions are clearly and logically drawn from and bounded by the data provided with no gaps in logic. ■ A reasonable and clear chain of logic from hypothesis to data to conclusions is made. ■ Conclusions attempt to discuss or explain conflicting or missing data.	■ Conclusions are completely justified by data. ■ Connections between hypothesis, data, and conclusions are comprehensive and persuasive. ■ Conclusions address and logically refute or explain conflicting data. ■ Synthesis of data in conclusion may generate new insights.

Criteria	Not Addressed	Novice	Intermediate	Exemplary
Discussion: Alternative Explanations and Limitations of Design	(0)	(1–2)	(3–4)	(5–6)
Alternative explanations are considered and clearly eliminated by data in a persuasive discussion. Limitations of the data and/or experimental design and corresponding implications are discussed.	■ Alternative explanations are not provided; ■ Limitations of the experiment and data are not discussed.	■ Alternative explanations are provided, but explanations include some trivial or irrelevant alternatives. ■ Limitations are discussed in a trivial way (e.g. “human error” is the major limitation invoked).	■ Reasonable alternative explanations are provided, but without relying upon data when possible. ■ Limitations are relevant, but conclusions fail to address or overstep the bounds indicated by the limitations.	■ Discussion of alternatives is reasonably complete, uses data where possible and results in at least some alternatives being persuasively dismissed. ■ Limitations are relevant and conclusions take these limitations into account, but do not undermine experimenters.
Discussion: Significance of Research	(0)	(1)	(2–3)	(4)
The paper gives a clear indication of the significance of the research and its future directions.	■ Significance and future directions are not addressed.	■ Significance and future directions are vague, implausible (not possible with current technologies or methodologies), trivial or off topic.	■ Significance and future directions are useful, but indicate incomplete knowledge of the field (e.g. suggest research that has already been done or is improbable with current methodologies). ■ Significance and future directions suggest a fruitful line of research, but lack detail to indicate motivations for or implications of the future research.	■ Significance and future directions are salient, plausible and insightful. ■ Significance and future directions suggest work that would fill knowledge gaps and move the field forward.

Criteria	Not Addressed	Novice	Intermediate	Exemplary
References	(0)	(1–4)	(5–6)	(7–8)
A search for reliable, relevant resources are evident. References are cited using CSE name-year format. Note: Proper format includes a one-to-one correspondence between in-text and end of text references (no references at end that are not in text and vice versa) and all in-text and end references are in CSE name-year citation style.	▪ References are not included or are not from acceptable or reliable sources (e.g. biased websites or unauthored blogs).	▪ References are limited. ▪ References include potentially unreliable sources and include limited primary sources. ▪ Citations are at least partially correctly cited and formatted.	▪ References are more extensive. ▪ References are used to provide some background information. ▪ References are likely reliable sources and include some primary sources. ▪ References are properly and accurately cited formatted with few mistakes.	▪ References indicate an extensive literature search was performed. ▪ References are used to provide ample background information. ▪ References are all very reliable sources and include sufficient primary sources. ▪ References are properly and accurately cited and formatted with no mistakes.
Writing Style	(0–1)	(2–5)	(6–8)	(9–10)
Grammar, word usage and organization facilitate the reader's understanding of the paper.	▪ Grammar and spelling errors significantly detract from the meaning of the paper. ▪ Word usage is frequently confused or incorrect. ▪ Subheadings are not used or are poorly used. ▪ Information is presented in a haphazard way.	▪ Grammar and spelling mistakes hinder the meaning of the paper. ▪ General word usage is appropriate, although use of technical language may have occasional mistakes. ▪ Subheadings are used and aid the reader somewhat. ▪ There is some evidence of an organizational strategy, though it may have gaps or repetitions.	▪ Grammar and spelling have few mistakes. ▪ Word usage is accurate and aids the reader's understanding. ▪ Distinct sections of the paper are delineated by informative subheadings. ▪ An organizational strategy is present with a logical progression of ideas.	▪ Correct grammar and spelling. ▪ Word usage facilitates the reader's understanding. ▪ Informative subheadings significantly aid reader's understanding. ▪ A clear organizational strategy is present with a logical progression of ideas. ▪ This paper is easier to read than most.

Guides to Writing

Writing Methods

The purpose of the Methods section is for you (the researcher) to let the reader know what you did. (See the appendix on lab reports for a rubric and more information about how you will earn a score.) Since you are letting the reader know what you did, this section is written in past tense and usually in passive voice. Methods are not written as a set of instructions; don't include a list (1, 2, 3, ... or first, second, third, ...). Even though you are not writing the Methods as a set of instructions, your Methods section should be repeatable. That means you'll need to include enough detail for someone to be able to repeat your experiment, while avoiding unneeded details. This may seem like a clear rule, but can still be difficult for new science students. Here's a non-science example that will illustrate the difference between too much unnecessary detail, and the right amount.

Let's say you run into a friend at the downtown library. He's new to town and asks how you got there from the Coliseum. Which of the following explanations would you give?

Explanation A

"To get to the library from the Coliseum, I grabbed my car keys out of my bag and walked to the lot where the car was parked. I used the key to enter the car and sit in the driver's seat. I started the ignition using the key, turning it to the right. Once the car was running, I put the car in drive (marked "D") using the gearshift and pulled out onto the Beechhurst Ave. heading south. I turned left onto Campus Dr. and proceeded to the light at the top of the hill. At the next intersection, I made a right onto University Ave. After going around the loop, but before passing the Mountainlair, I made a left onto College Ave., then very shortly after made a right to park into the Mountainlair parking garage."

Explanation B

"To get to the library from the Coliseum, I drove on Beechhurst Ave. heading south. I turned left onto Campus Dr. and proceeded to the light at the top of the hill. At the next intersection, I made a right onto University Ave. After going around the loop, but before passing the Mountainlair, I made a left onto College Ave., then very shortly after made a right to park into the Mountainlair parking garage."

Hopefully it is obvious that Explanation B would be more appropriate. Explanation A has way too much unnecessary information. There is no reason to explain how to enter and start your car, or put it into gear to drive. If your friend wanted to drive your route himself, he can figure out how to use his own car, but the explanation of the route you took would be helpful.

It might be more difficult to distinguish between necessary and unnecessary details when you are writing the Methods section of the experiment for a lab report. An analogy may be helpful. Here are excerpts from a Methods section. Which of these would be best?

> "Ten μL of blue dye was transferred into a blue microcentrifuge tube using a 2–20-μL pipette."

> "Ten μL of blue dye was transferred into a microcentrifuge tube."

Just like it wasn't necessary to explain how you started the car, it isn't necessary to explain how you used a 2–20-μL pipette to measure the volume or that your tube was blue, so the second one is better. You can assume that the reader has a basic understanding of lab equipment. If you simply gave the volumes and concentrations of solutions, your reader could repeat the experiment—he'd be able to pick which of his pipettes was the correct one to use, and he could put the dye in whatever color tubes he had available without affecting the results.

That explains how to omit irrelevant information, but what about enough information? Back to the example of getting from the Coliseum to the library:

> "I turned left, then right, then left, then right."

Clearly this one leaves out the essential street names that would enable someone to travel your route. Similarly, take a look at the science example:

> "Some dye was added to a tube."

Again, not nearly enough info is included; no one could repeat that part of the experiment.

As you proofread your Methods, continue to ask yourself if someone could repeat the experiment to make sure your level of detail is on target. In addition, don't forget to avoid lists and write in past tense. Also consider using passive voice to focus on the study rather than the researcher, and organize logically.

Creating Figures

Most scientific papers include tables and/or figures, such as graphs and images. Usually tables and figures are used to visually summarize results, but occasionally figures might be used to clarify or illustrate methods or experimental setup.

The first question students ask when they have an assignment that includes figures and tables is, "how many figures are required?" There is no single answer, but there are general guidelines to help you decide how many figures to include. In general, you should have a figure for each measurement. For example, let's say you conducted an experiment where you grew plants with different fertilizer treatments. You measured height, biomass, and number of flowers over time, so you could represent your data with three figures.

When presenting results using tables and figures, be sure that they are not redundant. Using the plant example, you wouldn't need to present average height in a table and a graph—just one of these would be sufficient. In addition, you usually want to present summarized data, like averages, rather that raw data. Raw data that represents each and every sample measurement isn't useful to the reader. Your job as the author of the paper is to summarize the data for the reader.

Figure legends, sometimes called captions, are a necessary part of each table and figure you present. They give important information that enables readers to quickly interpret the figure, such as what was tested, what was measured, and sample sizes. A figure with its legend should stand alone—it should be understandable without having to read the Results. While experienced readers of scientific literature have various preferences for how to approach reading papers, many skim through the Introductions, then jump straight to the Figures to get clues about what was studied and discovered. Then they go back and read from the beginning for more details.

With this in mind, consider the conventions about legends or captions as you create them. First, Tables and Figures are numbered consecutively in the order you refer to them in the text of the Results. In addition, legends include a title and a description of the figure:

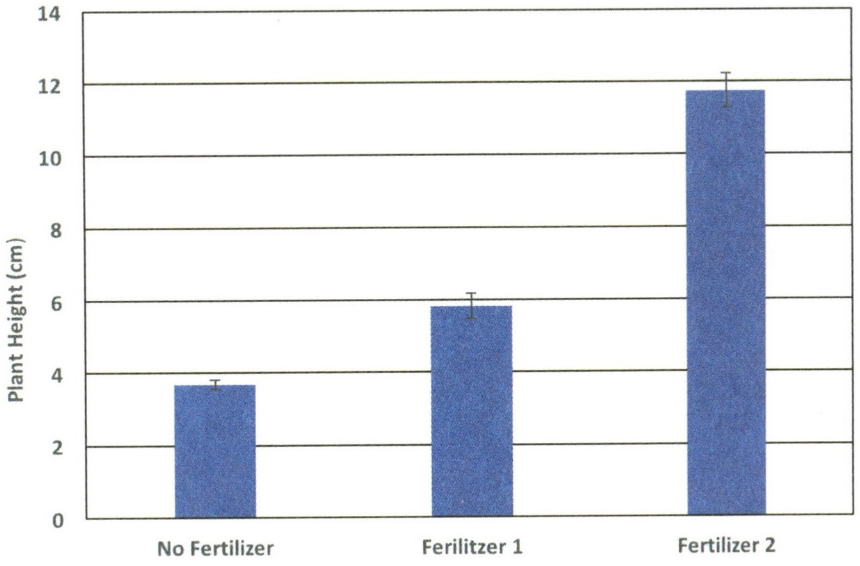

Figure 1 Effects of fertilizer treatment on plant height. Bars represent mean height of plants in various fertilizer treatments +/– s.e. ($n = 6$, $N = 18$).

There are conventions about legend placement too. For Figures, the legend goes below the figure, for tables, the legend goes above the table. An easy way to remember this is "table—top, figure—foot."

Writing Results

Results sections summarize your data trends in sentence form. Describe data trends even if you have presented them in figures or tables. Use parentheses to refer to a table or a figure:

"Plants in Treatment 1 had greater biomass at harvest (10 weeks) than plants in Treatment 2 (Figure 2)."

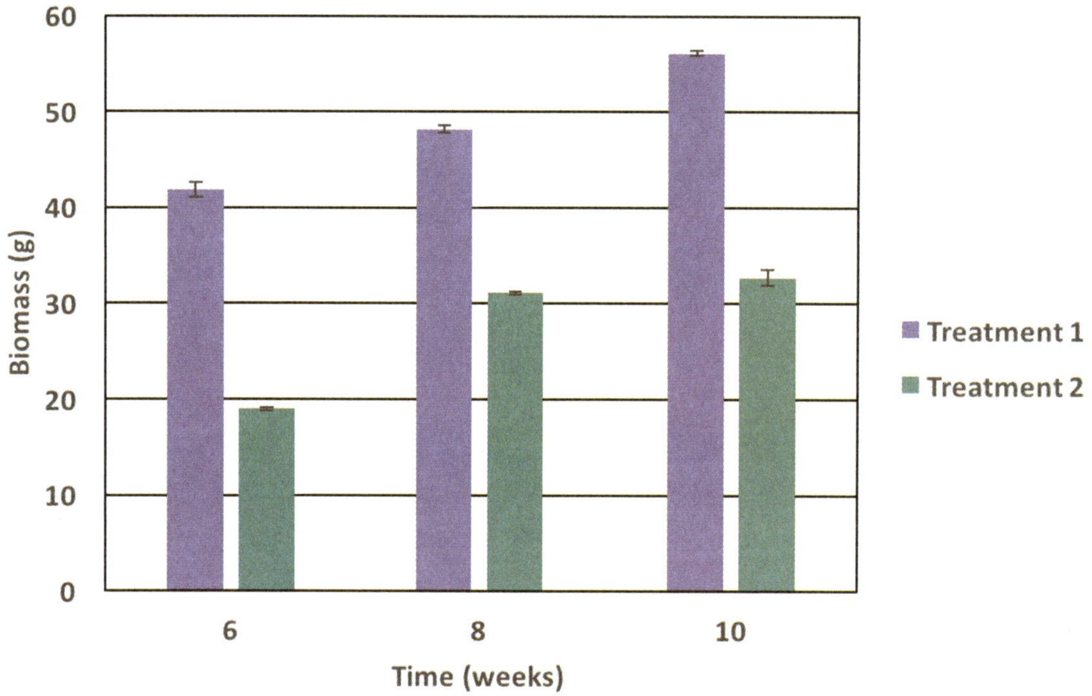

Figure 2 Treatment comparison of average biomass accumulation (+/– s.e.) at three time points. Treatment 1 and Treatment 2 effects on biomass were measured until harvest at 10 weeks, starting at 6 weeks post-seeding. Treatment 1 showed an average biomass accumulation of 42, 49, and 57 grams at 6, 8, and 10 weeks, respectively, while Treatment 2 showed a biomass accumulation at a lower rate at 18, 31, and 33 grams.

No explanation or interpretation is needed in the Results—save it for the Discussion. For example, it might be tempting to follow up the previous example sentence with "Plants in Treatment 1 were larger because …," but don't. Results are simply the facts.

For some lab reports, you will be instructed to use statistical analyses. If you have conducted statistical tests, such as a *t*-test, you will want to present the results of those tests in this section. Indicate *p*-values in parentheses. For example, "Plants grown with fertilizer 2 were significantly taller than plants grown with fertilizer 1 ($p = 0.002$)."

Low p-values indicate that there is a statistically significant difference between your treatment groups, while high p-values indicate there is no statistically significant difference. Be careful not to overstate the results. Sometimes it may appear that your treatment groups are different, but if your statistical tests don't fall below the threshold (usually $p < 0.05$), then you can't state there are statistically significant differences. For example, it is not okay to say, "The mean growth rate of group A was higher than group B. Group A had a growth rate of 1.2 cm per week, while group B had a growth rate of .98 cm per week ($p = 0.13$)." It is contradictory to indicate differences between treatment groups, unless the results of your statistical tests can support your statements.

Writing the Discussion

The Discussion section of the paper is probably one of the most difficult sections to write and requires some of the greatest skill. Logical organization will help you stay on track. Start this section with an indication of whether or not the hypothesis was supported. Then draw attention to the major trends in your data and provide an interpretation of the results. If your results are unexpected, you can suggest explanations. You may wish to point out limitations of the experimental design or execution. Indicate whether your results are consistent with others researchers. Finally, explain the implications of your research—explain how your findings are relevant to the bigger picture.

Acknowledging limitations of the experimental plan, methodology, or results is an important part of the discussion section. Limitations could include absence of previous data on the subject, inability to measure/collect/analyze data by optimal means, sample collection bias, or inability to test a larger sample size. By identifying sources of error or incompleteness of your data, you show that you have not only thought critically about the experimental design, but you have the expertise necessary to declare the potential impact this may have on your work and how it may change the interpretation of the data. If you point out weaknesses, you should also explain how they could be addressed by future studies or better methodology. When discussing limitations, be careful not to discredit yourself as a scientist or researcher. One of the biggest mistakes students make when writing the Discussion section is to blame flawed lab equipment or experimenter error as reasons for inconclusive or unexpected results, or an unsupported hypothesis. This just makes the reader question your competence, so don't betray yourself. Have confidence in your results without overstating conclusions beyond the limitations of your experiment.

Comparing your findings with work of other researchers can be one of the most challenging, but essential, parts of writing the Discussion. It is important in this section that you aren't just repeating results have already been stated in previous section of the paper. Instead should link your experiments to data from other researchers by referencing articles from the primary scientific literature. If others have found similar results, these references can provide additional support for your conclusions. However, don't be too dismayed if you find conflicting data in the literature. If you have come to a different conclusion, simply explain and support your interpretation. It may be helpful to contrast the studies in ways that could account for differences in findings, for example pointing out variation in methodology.

Even if you disagree with results from other scientists, be respectful of their work—explain why your conclusions differ based on facts, not just on personal opinions.

Below are examples of comparing and contrasting your studies with others:

> In agreement with other studies, our data supported the hypothesis that male rainbow trout *(Oncorhynchus mykiss),* a salmonine fish species that displays partial migration, rely on many factors that determine whether they migrate to spawn in fresh water or complete their entire life cycle in fresh water. Our data show diversity in the life histories of *O. Mykiss* that appears to be dependent on gene interactions with the environment and growth patterns of the fish. As also confirmed by Aubin-Horth et al. (2005), we found certain gene expression profiles related to growth pattern to be altered between male fish in the same populations who were either migratory or resident. However, in contrast with their data, we also found environmental factors play a role in whether an individual *O. mykiss* is migratory or resident. This is supported in other fish breeds by a study by Gross (1991) where the breeding behaviors of another salmonine fish, *Salmo salar* (Atlantic Salmon), have been largely affected over the course of the life of an individual salmon by environmental change.

Typically, the Discussion section includes a concluding paragraph. This paragraph helps connect your experiment to the broader impacts of the work. The results are stated briefly in context of the field, and how they impact the state of the research. In addition, future directions that you or other researchers can take to build upon the work are included. Experiments are not performed with the assumption that there is a defined end; there is always a next step.

Example of a concluding paragraph:

> In conclusion, we found migratory or residency patterns to be extremely diverse in various populations of *O. mykiss.* These patterns appear to rely on environmental factors, growth patterns, and genetic fitness landscapes and highlight the differences between male and female breeding behaviors. This study emphasizes the ability of species to respond to changing selective pressures, which obviously will have great consequence on not only *O. mykiss* and other salmonine species, but the entire ecological and evolutionary landscape of all inhabitants.

Writing the Introduction

It might seem counterintuitive that out of all of the parts of a lab report, you've been assigned the Introduction last. After all, in a scientific paper it is one of the first sections. However, from an author perspective, it makes sense to write the Introduction last. Here's why: after you have established your question and hypothesis, written the materials and methods, created figures and summarized results, and presented conclusions in the Discussion, you are in a position to put all of that information into perspective for the reader. Indeed, one of the goals of the Introduction is to entice someone to read your paper by alluding to what the paper contains. In addition, background information found in the Introduction serves to familiarize the reader with the topic, and references to the scientific literature provide context for the current experiment by letting the reader know what is already known. Combined, these provide a rationale for the entire experiment.

The Introduction of the paper should start off with the "big picture." Don't make the mistake of diving in too soon with a sentence like, "The purposes of this experiment was to …," or by jumping right to the hypothesis or predictions. Your reader won't yet have context for your experiment. Instead, think of the ideas presented in an Introduction being in a funnel shape—broad at the top and narrow at the bottom. For example, if you were doing an experiment involving a particular enzyme, say β-galactosidase, you could first indicate what the enzyme does and why it is important to living organisms.

> β-galactosidase is an enzyme that can catalyze the hydrolysis of galactose-containing carbohydrates into their component monosaccharides. In organisms that can metabolize lactose, β-galactosidase is responsible for breaking this disaccharide into the monosaccharides glucose and galactose. (Freeman et al. 2017)

Notice how the example above references a source for this background information. Some general background information can come from reliable secondary sources, like textbooks and review articles found in scientific journals. However, secondary sources referenced in your paper should be few and should **never include websites.** Once you broadly introduce the topic, you can add references from the scientific literature that support the rationale for your experiment. (Picture the funnel getting narrower.) Assume from the previous example that you compared β-galactosidase enzymes isolated from bacteria cells cultured under different conditions. You predicted that the rates of lactose breakdown would be different among the isolates. Here is an example of how to support your ideas with citations:

> Culture conditions of bacteria could affect the rates of lactose breakdown among β-galactosidase enzymes isolated from the different cultures. Scientists studying β-galactosidase enzymes purified from bacteria living in cold Antarctic seawater noted that the enzymes were active in cool temperatures. (Fernandez et al. 2002)

Once you have provided background information and built rationale for your experiment, you can end the Introduction with your question and hypothesis—you are at the narrowest part of the funnel. Although you've been practicing writing if-and-then style hypotheses in previous labs, this style is not a required format for the hypothesis in your lab report. However, you should present a complete scientific hypothesis—your reader should be able to pick out your possible answer to the question you've posed, they should have a general idea about how you are going to set up an experiment to test the hypothesis, and they should know what you expect to happen. All of these elements can be understood from the following example even though it isn't in if-and-then format.

> The effects of culture conditions of bacteria on the lactose hydrolysis activity of β-galactosidase were investigated. The enzymes isolated from bacteria cultured in high temperatures were expected to have higher activities than those cultured in low temperatures given than enzymes operate optimally under a narrow range of conditions, and this range may be influenced by the culture conditions under which the organisms are grown.

As you outline your Introduction, remember to keep the inverted funnel model for writing in mind. Start broad, tighten the scope, and end narrow.

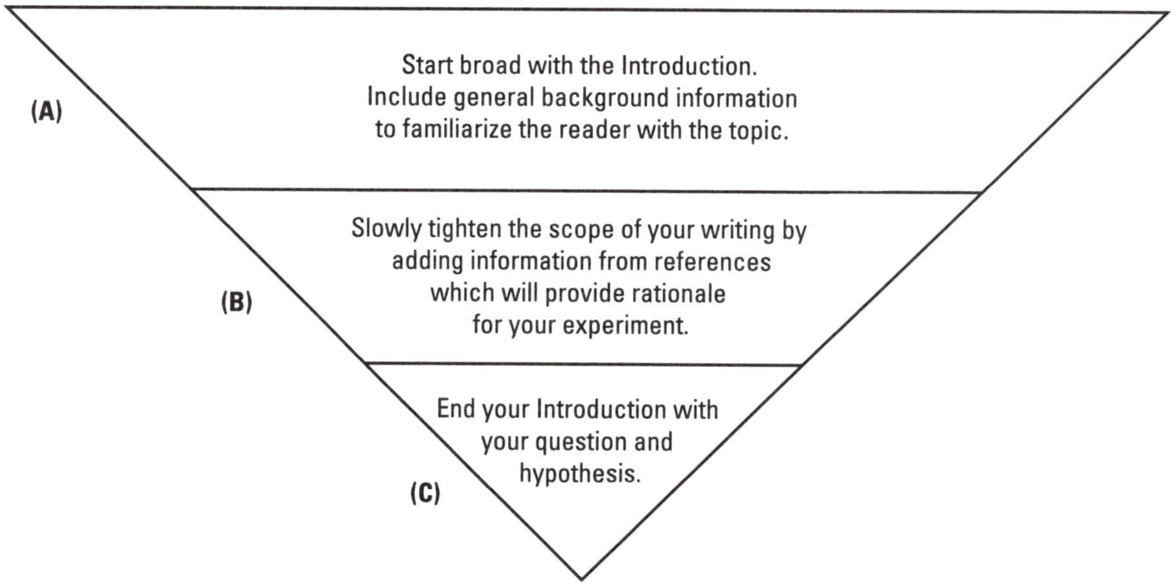

Figure 3

Writing Titles

After you've finished your writing assignment, you'll need to give it a title. Be concise, but descriptive. Also be sure to include all of the relevant information listed in the rubric: your name (the author), names of lab partners (your collaborators), name of your TA (your instructor), and course and section information.

Your readers' interest is captured by a good title that will let them know what to expect in the paper. A descriptive title will enable the readers to infer independent and dependent variables and even hint at the results of the experiment. Compare these two titles:

Title 1: "Plants and fertilizers."

Title 2: "Tomato plants grown with high-calcium fertilizers are less likely to develop blossom end rot on fruits."

The first title is not very descriptive. In contrast, the second title gives much more information. From the second title, you can infer the independent variable was fertilizers that differed in the amount of calcium. The dependent variable was likelihood of blossom end rot in fruits. You even know that plants grown in high-calcium fertilizers had less blossom end rot.

Review your titles before you submit and make sure that you include descriptive titles for all of your assignments.

Writing in Science

Tips on Writing in Science

- Don't rely on spell check. Many scientific words are not in pre-programmed dictionaries. Autocorrect may also suggest spellings or words that may be correct for some forms of writing, but not in the field. **Example:** "Absorbance" vs. "absorbent." Both spellings are correct, but the first indicates light transmittance in spectrophotometry, while the second indicates the ability of something to take in a substance, such as a sponge soaking up water.
- Include subheadings to help organize your writing within sections. For example, within a Methods section a paper may have a "Culturing fungi in-vitro" section followed by a "Measurements of fungal exudates" section.
- Know when to use numbers vs. numerals.
  - Spell out numbers ten or less (e.g., "… two species …").
  - Write numbers as numerals when greater than ten (e.g., "… 35 samples …") or associated with measurements (e.g., "10 µL" or "200 ng").
  - Spell out all numbers at the beginning of sentences or rearrange the sentence to avoid awkward wording (e.g., "Eighty seven thousand one hundred and ten new cases of melanoma are estimated …" Instead, try "An estimated 87,110 new cases …").
- Active voice is usually used in the main body of the paper, including the Introduction, Results, and Discussion, while passive voice is used in the Methods. For example, in the Introduction, an author might say, "We investigated three species of protected migratory birds." In the Methods, the author might say, "The migratory patterns of *Thalassarche melanophris, Phoebetria papebrata,* and *Diomeda exulans* were examined." Notice how the passive voice puts the focus on what's being studied rather than the person doing the study.

- Outlines will help you stay on track and help you make sure that each of your sections is important to the research question. Remember that your entire paper should follow an hourglass structure; start broad with the "big picture" at the beginning of the introduction, and then focus on your experiment at the end of the introduction with the hypothesis. The Methods and the Results will be a narrow focus on your experiment. In the Discussion, start narrow at the beginning to say whether or not the hypothesis was supported, then make it broad at the end when you put your results in context of the "big picture."
- "KISS" your writing—Keep it simple. Avoid flowery language with excess adjectives and adverbs. Use concise sentence structure and avoid run-on sentences. Scientific writing does not follow the same sentence structure and literary tropes as creative writing or other forms of communication. Make the "take home message" obvious.
- Don't write like you speak. Scientific writing is formal, so avoid slang terms or colloquial phrases. Scientific terms or jargon are often needed, but are only appropriate when they are defined or are commonly understood.

Scientific Species Names

- The species name is spelled out the first time it is introduced. This includes the genus and the species (**Example:** *Escherichia coli*). Sometimes a variant or subspecies is included (**Example:** *Brassica rapa* var. *glabra* or *Brassica rapa* subsp. *chinesis*). Notice that the "variant" and "subspecies" abbreviations are not italicized.
- Species names are always italicized. The first letter of the genus is always capitalized. The specific species name is all lowercase.
- Each subsequent time the species name is stated, it is abbreviated as the first letter of the genus, capitalized, and the species name, in lower case. **However,** if the species name is at the beginning of a sentence, **do not** abbreviate—write out the genus and species name. You can also rearrange the sentence to avoid the conflict.
- Common names are not capitalized.

Charles M, Kennedy S, Linda T, Moses M, Wonder N

Appendix B | Council of Science Editors (CSE) Name-Year Style for Formatting References

For this course, you will use the Council of Science Editors (CSE) name-year style for citing references in-text and in the of end-of-document Literature Cited or References section of your lab reports and assignments. Below are examples of how to format common types of sources that you will use.

Scientific Journal Articles

For the end references, list up to ten authors separated by commas. If there are more than ten authors, list the first ten, followed by "et al." For the in-text citation, list up to two authors, last names separated by "and." For more than two authors, list the first author's last name followed by "et al."

General Format

Author(s). Date. Article title. Journal title. Volume(issue):pages.

One Author

End reference:

Zambryski P. 1988. Basic process underlying *Agrobacterium*-mediated DNA transfer to plant cells. Annu Rev Genet. 22:1–30.

In-text citation:

… the T-DNA transfer process (Zambryski 1988)

Zambryski (1988) states that the T-DNA …

Zheng H Chen J Ahmad I

Two Authors

End reference:

Korb J, Foster KR. 2010. Ecological competition favours cooperation in termite societies. Ecol Lett. 13:754–760.

In-text citation:

… groups with internal strife (Korb and Foster 2010).

According to Korb and Foster (2010), groups with internal strife …

Three to Ten Authors

End reference:

Kim B, Du J, Eriksen DT, Zhao H. 2013. Combinatorial design of a highly efficient xylose-utilizing pathway in *Saccharomyces cerevisiae* for the production of cellulosic biofuels. Appl Environ Microbiol. 79(3):931–941.

In-text citation:

… taking advantage of random combinations (Kim et al. 2013).

According to Kim et al. (2013), pathway optimization …

Online Scientific Journal Articles

Use this online scientific journal article format if the article you are citing is an article that is found **exclusively** online. Many of the journal articles that you cite are available to you online through library websites and databases, but are also available in print; the online content is identical to the print content. If this is the case, go ahead and view it online, but reference it using the regular scientific journal format, which omits the accessed date, url, and doi. (Tip: If you see page numbers, it is probably available in print.)

General format:

Author(s) of article. Date of publication. Title of article. Title of journal (edition). [date updated; date accessed]; Volume(issue): page numbers. Location. Notes.

End reference:

Laranjiero R, Harinath G, Burke D, Braeckman BP, and Driscoll M. 2017. Single swim sessions in *C. elegans* induce key features of mammalian exercise. BMC Biol. [accessed 2017 May 23]; 15(30). https://bmcbiol.biomedcentral.com/articles/10.1186/s12915-017-0368-4. Doi: 10.1186/s12915-017-0368-4.

In-text citation:

(Laranjiero et al. 2017)

Books

Books are a secondary source of information, but may be useful as you complete assignments. For example, textbooks are an acceptable, reliable source of general information.

General format:

Author(s). Year. Title. Edition (if there is one). Place of Publication (City (State)): publisher.

End reference:

Morris JR, Hartl DL, Knoll AH, Lue R. 2013. Biology: how life works. New York (NY): W.H. Freeman & Company.

In-text citation:

(Morris et al. 2013)

End reference:

[CSE] Council of Science Editors, Style Manual Committee. 2014. Scientific style and format: the CSE manual for authors, editors, and publishers. Chicago (IL): Council of Science Editors in cooperation with the University of Chicago Press.

In-text citation:

(CSE 2014)

While the examples above represent the most common types of sources you'll cite for lab reports and assignments for this course, it is not a comprehensive list. If you find you need more information or examples on citations, the sources below may be useful:

- The CSE Manual for Authors, Editors, and Publishers, Scientific Style and Format Citation Quick Guide: *http://www.scientificstyleandformat.org/Tools/SSF-Citation-Quick-Guide.html*
- The Writing Center at the University of Wisconsin-Madison: *https://writing.wisc.edu/handbook/documentation/doccse/*
- WVU libraries, Citation Formatting: *http://libguides.wvu.edu/*